T0224572

SpringerBriefs in Electrical and Computer Engineering

Control, Automation and Robotics

Series Editors

Tamer Başar, Coordinated Science Laboratory, University of Illinois at
Urbana-Champaign, Urbana, IL, USA

Miroslav Krstic, La Jolla, CA, USA

SpringerBriefs in Control, Automation and Robotics presents concise summaries of theoretical research and practical applications. Featuring compact, authored volumes of 50 to 125 pages, the series covers a range of research, report and instructional content. Typical topics might include:

- a timely report of state-of-the art analytical techniques;
- a bridge between new research results published in journal articles and a contextual literature review;
- a novel development in control theory or state-of-the-art development in robotics;
- an in-depth case study or application example;
- a presentation of core concepts that students must understand in order to make independent contributions; or
- a summation/expansion of material presented at a recent workshop, symposium or keynote address.

SpringerBriefs in Control, Automation and Robotics allows authors to present their ideas and readers to absorb them with minimal time investment, and are published as part of Springer's e-Book collection, with millions of users worldwide. In addition, Briefs are available for individual print and electronic purchase.

Springer Briefs in a nutshell

- 50–125 published pages, including all tables, figures, and references;
- softcover binding;
- publication within 9–12 weeks after acceptance of complete manuscript;
- copyright is retained by author;
- authored titles only – no contributed titles; and
- versions in print, eBook, and MyCopy.

Indexed by Engineering Index.

Publishing Ethics: Researchers should conduct their research from research proposal to publication in line with best practices and codes of conduct of relevant professional bodies and/or national and international regulatory bodies. For more details on individual ethics matters please see: https://www.springer.com/gp/authors-editors/journal-author/journal-author-helpdesk/publishing-ethics/14214

Wouter Jongeneel · Emmanuel Moulay

Topological Obstructions to Stability and Stabilization

History, Recent Advances and Open Problems

 Springer

Wouter Jongeneel
Risk Analytics and Optimization Chair
(RAO)
École Polytechnique Fédérale de Lausanne
Lausanne, Switzerland

Emmanuel Moulay
XLIM Research Institute (UMR CNRS
7252)
University of Poitiers
Poitiers, France

The open access publication of this book has been published with the support of the Swiss
National Science Foundation.

ISSN 2191-8112   ISSN 2191-8120 (electronic)
SpringerBriefs in Electrical and Computer Engineering
ISSN 2192-6786   ISSN 2192-6794 (electronic)
SpringerBriefs in Control, Automation and Robotics
ISBN 978-3-031-30132-2   ISBN 978-3-031-30133-9 (eBook)
https://doi.org/10.1007/978-3-031-30133-9

This Springer imprint is published by the registered company Springer Nature Switzerland AG
The registered company address is: Gewerbestrasse 11, 6330 Cham, Switzerland

Preface

This book showcases the unsurpassed effectiveness of employing topological methods in studying the stabilization of nonlinear dynamical systems. Strikingly, one observes a stark difference in the ramifications of controllability for either linear or *non*linear systems. For linear dynamical systems, controllability implies the existence of a *continuous, globally asymptotically stabilizing* feedback. On the other hand, for controllable nonlinear dynamical systems, there might not exist a feedback that is both continuous, globally well-defined and stabilizing in the asymptotic sense. This can be understood from the topological viewpoint.

A thorough understanding of this phenomenon is of great importance as control theoretic tools are being applied to systems of ever-growing complexity whilst demanding increasingly strong guarantees, e.g., in the context of algorithm design, ecology, reinforcement learning, robotics and so forth. With the increase in applications—especially the ones outside of the traditional control domain—comes the responsibility of precisely detailing what the tools can and cannot do. Failing to do so can have catastrophic consequences, e.g., incorrectly assuming that global stabilization is possible based on merely numerical simulations. In other words, given a control system, can some desirable dynamical behaviour possibly be prescribed? Here, the desirable behaviour can oftentimes be captured by the *stability* or *stabilization* of some set. For instance, the stationary pumping of the heart and the gait of a walking robot can be captured by the stability and stabilization of a periodic orbit, respectively.

Ever since the time and work of Poincaré, it has been known that one can greatly benefit from the *qualitative* viewpoint in studying such a question of admissible stable behaviour. Doing so, topological methods present themselves as natural methodological candidates as these tools allow for drawing strong qualitative conclusions based on merely elementary structural knowledge of the control problem at hand. This study took off in the second half of the 1900s and rapidly brought about a vast amount of work in the intersection of control theory, dynamical systems and topology. To this day, this line of work provides for generic and unique insights in theoretical and applied control, e.g., necessary conditions for continuous asymptotically stabilizing feedback controllers to exist. Unfortunately, most of the results are not as widely known as desirable. Hence, motivated by this observation and the theoretical and applied prospects, we believe this is the right time to provide the first unified and complete overview of topological methods in studying nonlinear stability and stabilization.

Specifically, this monograph aims to provide a self-contained overview accessible to the graduate student interested in control theory. In particular, this work aims to provide a unified overview of topological obstructions to stability and stabilization of dynamical systems defined on topological spaces. We review the interplay between the topology of an attractor, its domain of attraction and the underlying topological space, e.g., a manifold, that is supposed to contain these sets. Some proofs of known results are presented to highlight assumptions and to develop extensions. To be as complete as possible, we also provide a few new results and we highlight the most popular and effective methods regarding how one can overcome these topological obstructions. Moreover, we show how Borsuk's retraction theory and the index theoretic methodology by Krasnosel'skiĭ and Zabreĭko underlie a large fraction of results known today. This point of view naturally reveals important open problems and as such we believe that this work could be of interest to any student and researcher in control theory, dynamical systems, topology or a related field.

Lausanne, Switzerland   Wouter Jongeneel
Poitiers, France   Emmanuel Moulay
February 2023

Acknowledgements

The preparation of this publication was supported by the Swiss National Science Foundation. WJ is supported by the Swiss National Science Foundation under the NCCR Automation, grant agreement 51NF40_180545, and would like to thank Nicolas Boumal, Timm Faulwasser, Daniel Kuhn, Isabelle Schneider and Babette de Wolff, their feedback greatly improved the work. EM would like to thank Emmanuel Bernuau, Sanjay P. Bhat, Patrick Coirault and Qing Hui for many discussions and feedback on topological obstructions to stability and stabilization. Both authors would like to thank Oliver Jackson for his help throughout the publication process.

All figures are made using Inkscape with TexText.[1]

---

[1] See https://inkscape.org/ and https://textext.github.io/textext/.

Contents

1 Introduction 1
   1.1 Impetus 1
   1.2 Historical Remarks 4
        1.2.1 Topology 4
        1.2.2 Dynamical Systems 6
        1.2.3 Modern Control Theory 8
   1.3 Case Study: Optimal Control on Lie Groups 10
   1.4 Content and Structure 12
   References 13

2 General Topology 21
   2.1 Topological Spaces 21
   2.2 Homotopy and Retractions 23
   2.3 Comments on Triangulation 25
   References 26

3 Differential Topology 29
   3.1 Differentiable Structures 29
   3.2 Submanifolds and Transversality 29
   3.3 Bundles 31
   3.4 Intersection and Index Theory 33
   3.5 Poincaré–Hopf and the Bobylev–Krasnosel'skiĭ theorem 40
   References 45

4 Algebraic Topology 49
   4.1 Singular Homology 49
   4.2 The Euler Characteristic 52
   References 54

5 Dynamical Control Systems 57
   5.1 Dynamical Systems 57
   5.2 Lyapunov Stability Theory 64

5.3   Control Systems 65
References 72

6   Topological Obstructions 77
6.1   Obstructions to the Stabilization of Points 78
6.1.1   Local Obstructions 79
6.1.2   Global Obstructions 84
6.1.3   A Local Odd-Number Obstruction to Multistabilization 90
6.2   Obstructions to the Stabilization of Submanifolds 93
6.3   Obstructions to the Stabilization of Sets 98
6.4   Other Obstructions 102
References 103

7   Towards Accepting and Overcoming Topological Obstructions 109
7.1   On Accepting the Obstruction 109
7.2   On Time-Varying Feedback 110
7.3   On Discontinuous Control 111
7.3.1   Hybrid Control Exemplified 111
7.3.2   Topological Perplexity 114
References 114

8   Generalizations and Open Problems 119
8.1   Comments on Discrete-Time Systems and Periodic Orbits 119
8.2   Comments on Generalized Poincaré–Hopf Theory 120
8.3   A Decomposition Through Morse Theory 120
8.4   An Application of Lusternik–Schnirelmann Theory 121
8.5   Introduction to Conley Index Theory 122
8.6   Conclusion and Open Problems 124
References 127

Series Editor Biographies 131

Chapter 1
Introduction

1.1 Impetus

From climate models to walking robots and from black holes to the economy; many objects of science are studied by means of *dynamical systems* evolving on *manifolds* [40, 105, 151, 156]. As models are not perfect and explicit solutions are rare, ever since the time of Poincaré the interest shifted from studying the *quantitative-* to studying the *qualitative* behaviour of a dynamical systems at hand. Not only the *description* of a system, but in particular, the *prescription* of the dynamics of a system became of increasing importance. Naturally, one must ask if the desirable dynamics are admissible in the first place. Topology provides for a rich set of answers relying on a minimal set of assumptions, as surveyed in this work.

Given some space M and some subset A of M. We will be mostly concerned with studying if M admits a dynamical system such that the set A, e.g., some configuration of a robot, is (uniformly) globally *asymptotically stable*. This stability notion is captured by: (i) *Lyapunov stability*: that is, for each neighbourhood U of A there is another neighbourhood $V \subseteq U$ of A such that when the system starts from a state within V, the state of the system will stay in U; and (ii) *attractivity*: that is, there is a neighbourhood W of A such that when the system is started from a state within W, the state of the system converges asymptotically to A. When $W = $ M, we speak of *global* asymptotic stability. We also remark that Lyapunov stability is sometimes referred to as simply *stability*, consequently, a set that fails to be stable, with respect to some dynamical system, is said to be *unstable*. For formal definitions, see Chap. 5.

Example 1 (*Admissible flows on the circle*) Let one be tasked with finding a continuous flow, i.e., a map that defines state propagation as a continuous function of the time to propagate and the instantaneous state, such that some point $p^\star$ on the circle $\mathbb{S}^1$ is globally asymptotically stable, e.g., see Fig. 1.1(i). The flow in any small neighbourhood around $p^\star$ is well-defined, but one eventually runs into problems, see Fig. 1.1(ii), and cutting the circle (allowing for discontinuities) seems the only solution, see Fig. 1.1(iii). Although, relaxing the task, e.g., by allowing for *almost surely* global asymptotic stability, *local* asymptotic stability or merely global *attractivitiy* (no Lyapunov stability), also belongs to the possibilities, see Fig. 1.1(iv)–(vi). In fact, by studying Fig. 1.1 one might observe some patterns, e.g., stable and unstable

© The Author(s) 2023
W. Jongeneel and E. Moulay, *Topological Obstructions to Stability and Stabilization*,
SpringerBriefs in Control, Automation and Robotics,
https://doi.org/10.1007/978-3-031-30133-9_1

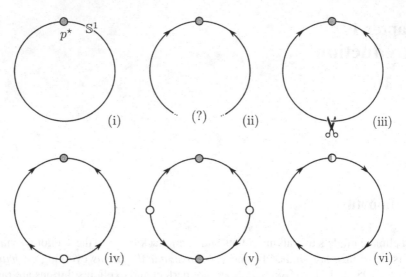

Fig. 1.1 Example 1, (asymptotically) stabilizing $p^\star \in \mathbb{S}^1$

equilibrium points necessarily come in pairs. The key observation, however, is that for compact nonlinear spaces, the study of global behaviour should take the global *topology* into account.

Example 1 illustrates the fact that merely the underlying *topology* of a space can obstruct the existence of certain *qualitative* behaviour.

Next, going one step beyond the circle, we consider the mathematical pendulum, cf. [142, 145], which displays a myriad of topological phenomena [97, 126] and captures the dynamics integral to the study of robotics, aerial vehicles and more.

Example 2 (The mathematical pendulum) The single-link pendulum displays intricate nonlinear behaviour by having the circle $\mathbb{S}^1$ as its *configuration space*. As a pendulum is a second-order system, the state space, however, becomes the cylinder $\mathbb{S}^1 \times \mathbb{R}$, parametrizing the angle and rotational velocity, or as will be discussed, the trivial vector bundle $\pi : \mathbb{S}^1 \times \mathbb{R} \to \mathbb{S}^1$. We assume that one can control the pendulum by means of a torque applied to its axis and that this torque is chosen as a continuous function of the state. Moreover, we assume that this feedback, i.e., the torque as a function of the state, gives rise to a *continuous* flow on $\mathbb{S}^1 \times \mathbb{R}$ and that there are isolated fixed points of the flow, e.g., we assume there is some form of friction. Now we ask a similar question as before, can the pendulum be globally asymptotically stabilized in the upright ($\theta = 0$) position by such a feedback? This is a *stabilization* problem. For example, Fig. 1.2 displays trajectories analogous to Fig. 1.1(iv)–(v). The reader is invited to construct the phase portrait akin to Fig. 1.1(vi) and recover what is called the *unwinding phenomenon*. One observes that generalizing the circle to the *non*-compact cylinder did not improve the situation, *topological obstructions* prevail. Simultaneously, this example shows the power of this line of study in that

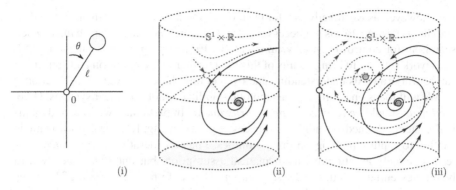

Fig. 1.2 Example 2, asymptotically stabilizing the inverted pendulum

the results are general; we did not yet make any explicit modelling assumptions, e.g., regarding friction and inertia.

Example 2 sketches the practical value of studying topological obstructions; when explicit models are unknown or too uncertain, the underlying topology can already provide insights in admissible *qualitative* behaviour. In fact, one can argue that topology is the natural language to study these kind of dynamical problems [83].

The previous examples focus on equilibrium *points*, however, in many applications one might be interested in stabilizing non-trivial periodic orbits or other *sets*. The intuition from before can be generalized, for example, consider globally asymptotically stabilizing the circle $\mathbb{S}^1$ in the plane $\mathbb{R}^2$. Again, obstructions of this kind are purely topological.

Although results of this nature go back to the 1800s, we feel there is a need to survey existing material: the ever-growing field of applied dynamical systems theory, including, but not limited to, motion planning, numerical optimal control, system identification and reinforcement learning. The aforementioned observations have particularly important ramifications in those areas, as frequently, one needs to specify a space of models or policies to optimize over, a priori.

As stressed in a recent article by Schoukens and Ljung [136], nonlinear system identification poses challenges beyond linear system identification, in particular, the nonlinear structure prohibits straight-forward extrapolation, that is, inferring global information from local data is inherently difficult in the nonlinear regime. However, knowledge of the underlying topological space *is* occasionally present, as such it is worthwhile to study ramifications of just the topological data at hand.

Regarding policies, it is important to highlight that controllability merely implies the *existence* of *some* admissible input "*steering point A to point B*". As stressed by Sussmann [146, p. 41], only when the admissible inputs are precisely detailed, one can study if some control objective can be satisfied by selecting the input as some kind of feedback controller. In practice, when numerically optimizing over policies, one is for example drawn to employing some form of function approximators [29]. It might be tempting to believe that the space of continuous functions is sufficiently rich, however, as we already saw for the most elementary nonlinear manifolds, when stability is desired, this is not necessarily true. A similar argument can be made when searching for control-Lyapunov functions (CLF) or control barrier functions [88]. This work sets out to bring results of this kind further to the attention and spur more future work towards understanding and overcoming these topological obstructions.

We will focus on continuous asymptotic stabilization of nonlinear systems. Here, *nonlinear* should be read as *not necessarily linear*. In particular, we look at dynamical systems defined on nonlinear spaces, e.g., as in Fig. 1.1, local linearizations can fail to capture topological impossibilities. The consideration of *dynamical* systems is largely an attractive mathematical assumption, but one that is believed to be quintessential to better understand larger classes of physical systems. The focus on *continuity* is historically based on implementation and robustness considerations and seems at first a general assumption. Enforcing continuity allows for a better understanding of how general this assumption really is. The focus on continuity also puts work on neural networks in perspective as common architectures result in maps which are at least continuous. The desire of (uniform) *asymptotic stabilization* is a natural one from the classical mechanics point of view and enviable with respect to robustness, but also here one will observe that this demand can be too strong.

1.2 Historical Remarks

The study of topological obstructions in the context of dynamical control systems is at its core the investigation of what kind of—qualitative speaking—dynamical systems a space admits. Philosophically, this is in line with the early work on topology and dynamical systems as pioneered by Poincaré. To put the material in perspective, the next section briefly covers this history at large.

1.2.1 Topology

The fourth axiom of Euclid states *"Things which coincide with one another are equal to one another"* [4, p. 6]. Although Euclid was a geometer and no topologist, this axiom is broadly stating what topology would be all about. Yet, *"things"* and *"equivalence"* had to come a long way since the time of Euclid.

In the early 1900s, Cantor started the development of set theory and contributed to the initial work on topology [31]. When Cantor his one-to-one map from the interval to the hypercube revealed the intricacies of defining *dimension*,[1] it was Dedekind to point out that perhaps something is missing: *continuity* [112]. Peano's space-filling curve showed that even when continuity is satisfied (but injectivity is lost), counter-

---

[1] Cantor famously wrote *"I see it, but I don't believe it"* to Dedekind. See [49] for more context.

intuitive phenomena can still be observed [115]. These *counterexamples* revealed a lack of understanding when it comes to classifying objects as being "*equivalent*".

Poincaré was amongst the first to define, in for example his 1895 "*Analysis situs*", what this equivalence could be: a continuous one-to-one transformation, called a *homeomorphism* [121]. It took, however, a while before a homeomorphism meant what it does today. Poincaré described *analysis situs* (the predecessor of *topology*, attributed to Listing) as "*This geometry is purely qualitative; its theorems would remain true if the figures, instead of being exact, were roughly imitated by a child.*" [123]. A weaker invariant that plays a substantial role in this work is that of *homotopy*. It can be argued that homotopies, albeit with fixed endpoints, originated in the work of Lagrange on the calculus of variations [41]. Although homotopies (described as continuous deformations), the fundamental group and initial homology theory appeared in the work of Poincaré [121], the formal description of what this "*continuous deformation*" is supposed to be, was missing. By building upon Schönflies, the concept of *dimension* that bothered Cantor, was eventually put on a theoretical footing by Urysohn [152] and most notably by Brouwer [25, 26] around 1910, see also [72, Chap. 1]. This work by Brouwer also formalized homotopies and their equivalence classes as we know them today [27]. Besides, it brought forward the concept of *degree*, a notion of importance in this work, yet, a notion that was to some extent already known to Cauchy, Picard and in particular Kronecker [43, 114, 137]. Hurewicz added to this line of work by defining when spaces are homotopic [68] and a related theory, that of *retractions* was pioneered by Borsuk [16] in the late 1920s. A concept intimately related to the degree of a map is that of an *index* of a vector field, as arguably introduced by Poincaré and further developed by Hopf [66, 67]. An important elaboration and formalization is due to Brouwer [42, p. 168], Poincaré often assumed differentiability or even analyticity of objects under consideration [72, p. 57], while Brouwer relaxed this to mere continuity and was able to formally apply index theory to continuous vector fields on the sphere [26].

In the meantime topology branched out. Although the "*Euler characteristic*" was known, it can be argued that Riemann founded *algebraic* topology [17, pp. 162–164], as, amongst other things, he evoked, what would be called *Betti numbers*, in the late 1800s with his study into connectedness [130], [72, Chap. 2]. Generalizations required work from Betti [9], Poincaré [121] and most importantly, formalizations by Noether[2] [111] and later Eilenberg [45].

Formalizing the abstract study of sets relied on early work by Hilbert [61], Fréchet [46], Riesz [131] and in particular Hausdorff, whom in 1914 published *Grundzüge der Mengenlehre* [56]. This was the first axiomatic work on abstract topological spaces and can be seen as the start of *general* topology known today. Hausdorff laid down the foundation of (general) topology and provided the neighbourhood generalization of the Bolzano–Cauchy $\varepsilon - \delta$ continuity definition, although the neighbourhood concept was already known to some extent.

---

[2] For more context on Noether her contributions, e.g., in relation to pioneering work by Vietoris and Alexandroff & Hopf, see [64].

Concurrently, the notion of a (linear) *manifold* was already known to Gauss, but, amongst others, it were Möbius [106], Jordan [74] and in particular Riemann [129] and Poincaré [121] that initiated the classification of manifolds. The theory of differentiable manifolds, however, and thereby *differential* topology was largely developed by Weyl [72, Chap. 2], Veblen and Whitehead (J.H.C.) [154] and Whitney [157, 158], e.g., this includes formalizations of *coordinate charts, tangent spaces* and *embeddings*. The initial work by Cartan (Élie) on *fiber bundles* was further developed by Seifert, Whitney and Ehresmann [72, Chap. 22]. A later, but instrumental contribution for this work is the notion of *transversality* as developed largely by Thom [149]. The most important development for this work, and perhaps one of the most important series of results in the intersection of topology and dynamical systems in general, is the *Poincaré–Hopf theorem*, with contributions by Gauss, Kronecker, Bonnet, Dyck, Brouwer, Poincaré and in particular Lefschetz [93, 94] and Hopf [67], see also [48].

For these and more examples, see [42, 72, 137, 148, 155], the historical notes in [17, 43, 114, 163] and [124] for the English translation of [121]. Also, as highlighted, topology thrived on counterexamples, which remained an active area [143].

1.2.2   Dynamical Systems

Starting with Newton, the study of dynamical systems was dominated in the early days by the *quantitative* study of stability in the context of celestial mechanics. The complications that arise when trying to explicitly solve differential equations were early understood and amongst others, Laplace, Lagrange, Poisson and Dirichlet all claimed to have proven that the solar system was stable by means of analyzing series expansions, e.g., see the introduction in [1]. Motivated in part by a competition in the late 1880s hosted by Oscar II, the king of Sweden & Norway, it was Poincaré who pointed out that the series expansion approach of that time was flawed. As he puts it, *"There is a sort of misunderstanding between geometers [mathematicians] and astronomers about the meaning of the word convergence." "...take a simple example, consider the two series,*

$$\sum_n \frac{1000^n}{n!} \quad \text{and} \quad \sum_n \frac{n!}{1000^n},$$

"geometers would say the first series converges, ...but they will regard the second as divergent. "Astronomers, on the contrary, will regard the first series as divergent as the first 1000 terms are increasing and the second as convergent as the first 1000 terms are decreasing" [120, pp. 1–2]. Poincaré emphasized throughout his work that divergent series could have practical value, but to prove anything meaningful about the stability of the solar system required a rigorous mathematical justification [120, p. 2]. It was during this time that Poincaré developed his qualitative methods for mechanics and dynamical systems, e.g., [118–120, 122]. Amongst other things, in this work he started the classification of qualitative dynamical systems, introduced

the notion of the vector field index and advocated the use of transversality (Poincaré sections), topics we return to below. Ironically, the original version of the award winning[3] 1890 paper [118] did not contain the most celebrated parts of the work, e.g., work on homoclinic solutions (initially called *doubly asymptotic* solutions), only the corrected version did. Upon fixing the errors in the initial version of [118], Poincaré was puzzled by the chaos he created (found) and wrote to Mittag-Leffler *"...I can do no more than to confess my confusion to a friend as loyal as you. I will write to you at length when I can see things more clearly."* [5, Sect. 5.8]. Almost ten years later, the situation was far from clarified, regarding homoclinic solutions to the three-body problem, Poincaré writes *"One is struck by the complexity of this figure that I am not even attempting to draw."* [122, p. 389]. It would take more than half a century before Smale would clarify the situation.

As remarked previously, one of the main topics of study has always been that of *stability*. Around the same time as the early work of Poincaré, Lyapunov (Liapunov)— who was inspired by Poincaré [102, pp. 531–532]—published his thesis on qualitative stability theory in the early 1890s [99]. Although Lyapunov was the first to lay down the foundations, similar notions appeared in the work of Lagrange [90] and Dirichlet [44]. In this work, Lyapunov devised two lines of attack to reason about (local) stability. The first method (*indirect*) relies on linearizing the dynamics, whereas the second method (*direct*) is in the spirit of the work by Lagrange and looks for an "*energy*" function (Lyapunov function), which is strictly positive, yet decreasing along the dynamics. Attributed to Poission, Poincaré, however, spoke of stability of a point when a trajectory returned infinitely often to points arbitrarily close to where the point started from. Lyapunov spoke of stability of a point when for each open neighbourhood U around a point there is an open set $V \subseteq U$ such that each trajectory starting in V remains in U. Lyapunov's approach to stability (the second, direct method) had an intuitive and almost direct link to modelling paradigms in mechanics (energy) and it grew out to be one of the most celebrated tools in the study of dynamical systems cf. [10, 81, 89].

Concurrently, after proving Poincaré's "*last geometric theorem*" in 1913 [11], it was in particular Birkhoff who propelled the qualitative study of dynamical systems [12]. This qualitative viewpoint brought the general notion of stability more to the forefront, not only stability of a system, but also the stability of its description, called *structural stability*. Structural stability for 2-dimensional systems was introduced by Andronov and Pontryagin [2] and extended to 2-dimensional manifolds by Peixoto [116]. Smale is largely responsible for further abstractions and continuation of this line of work [139]. In particular, his creation of the horseshoe-map clarified what Poincaré was having trouble with in his work on the three-body problem: the intricate dynamics close to a homoclinic equilibrium point. Interestingly, Smale

---

[3] With Hermite, Mittag-Leffler and Weierstrass in the jury.

also made a significant contribution to another, yet topological, open problem by Poincaré, he proved the (generalized) Poincaré conjecture[4] for $n \geq 5$ [138].

In passing, we highlight a few other influential works, contributing to (the early development of) the qualitative theory of dynamical systems. The work by Hadamard [53, 54]—who was, interestingly, in close contact with Brouwer, Lefschetz [95], LaSalle (La Salle) [91] and Hartman [55]. Early contributions to stability theory by Hermite, Routh and Hurwitz, e.g., see [69, 77]. The initial work on center-manifold theory by Pliss [117] and Kelly [80]. Catastrophe theory by Thom [150]. Converse Lyapunov theorems and its topological ramifications due to Kurzweil [87], Bhatia and Szegö [10] and Wilson [159, 160]. The concept of a region of attraction due to Aĭzermann, Barbašin, Krasovskiĭ, Nemyckiĭ and Stepanov, e.g., see [10, 86]. The work by Kolmogorov, Arnol'd and Moser, e.g., see [24]. The work by Hirsch et al. [63]. The topological critical-point theory by Morse [107] and the work on chaos theory by Takens and Ruelle [132].

For more information, see [34, 48, 62, 65], or the historical notes in [1, 10, 50]. In particular, see [5] for an exposition of the competition organized by Mittag-Leffler, the initial error in the work submitted by Poincaré, how this was resolved and how the mathematical community responded. For more on the history of stability, see [77, 96]. See [125], for the English translation of [118].

1.2.3 Modern Control Theory

In the late 1700s, the field of control theory emerged due to a growing practical interest in improving the performance of mechanical systems. As discussed above, the 1800s gave rise to a lot of theoretical work on describing the dynamics of a system and in particular studying its stability. Nevertheless, motivated by the needs of war and after original work on telecommunication, filtering and circuit design in the frequency-domain, *modern* control theory, however, was only born in the mid 1900s out of the pioneering work by Kalman[5] [78, 79], Bellman [6], Pontryagin [47, 127] and their coworkers. This line of work emphasized some benefits of the *state space* approach (the time-domain) and essentially reconnected control theory to the early work of Poincaré and Lyapunov. The state space approach to linear control theory also brought linear algebra more to the forefront, which opened the door for a rigorous approach to nonlinear control, not merely by approximation, but also by appealing to differential geometric tools, cf. [21, 70, 110, 161]. Perhaps the central topic of study in (deterministic) control theory in the late 1900s was that of controllability, i.e., all questions related to the possibility of steering a system "*from A to B*". Naturally,

---

[4] This conjecture states (in the category of topological manifolds): every closed n-dimensional manifold homotopy equivalent to $\mathbb{S}^n$ is homeomorphic to $\mathbb{S}^n$. This conjecture was partially open until 2003, when Perelman filled in the gap for $n = 3$, see [144] for more on this program.

[5] Lefschetz came out of retirement in the late 1950s to start a group in Baltimore on nonlinear differential equations and no other than Kalman started his professional career in this group.

these questions relate to the aforementioned work on dynamical systems, e.g., if a space does not admit a dynamical system with a certain property, then clearly no input exists that can enforce it. Building upon the work of Chow[6] in 1940 [33], it can be argued that work on nonlinear controllability started in the 1960s—just after Kalman published his work on linear controllability—with influential contributions by Hermann [58], Lobry [100, 101], Haynes and Hermes [57], Sussmann and Jurdjevic [147], Brockett [20] and most notably Hermann and Krener [59].

In 1978, Jurdjevic and Quinn constructed a controllable system on $\mathbb{R}^2$ that cannot be stabilized via differentiable feedback [75]. Then, against to what was a common belief at the time, by constructing an example on $\mathbb{R}^2$, Sussmann showed in 1979 that controllability does, however, also not implies that a stabilizing *continuous* feedback exists [146]. A year later, Sontag and Sussmann developed theory underpinning scalar examples along these lines [140]. These examples were not unparalleled as in 1983 Brockett gave an explicit necessary (topological) condition for stabilizing differentiable feedback laws to exist [23]. Brockett's condition gave rise to many examples, as a lot of controllable systems failed to adhere to this condition. This, and earlier work by Kurzweil [87], Wilson [160] and Bhatia and Szegö [10] can be seen as a start of the work on topological obstructions to stability and stabilization.

For more on the development of nonlinear controllability and related tools see [32, 71, 98, 153]. For more on the history of control theory, see [13] and for a historical account by Brockett himself, see [19]. See also [22, 60] for early works by Brockett and Hermann & Martin, respectively, highlighting the use of a topological and geometrical viewpoint in the context of system and control theory.

At last, we emphasize two additional schools. First, in the East, Krasnosel'skiĭ and coworkers elaborated during the second half of the 20th century on a blend of most of the aforementioned material in their study of topological methods in nonlinear analysis [84, 85]. As will be discussed below, the monograph by Krasnosel'skiĭ and Zabreĭko contains a variety of results related to arguably the most influential control-theoretic topological result produced in the West—known as *"Brockett's condition"*, as discussed in detail in Chap. 6 cf. [84, Chaps. 7–8], [23]. As also pointed out in [113], although the translated version of their monograph appeared in 1984, the original Russian version appeared in 1975, well before that particular work by Brockett. Moreover, Krasnosel'skiĭ's earlier monograph from 1968 [85] and a 1974 paper by Bobylev and Krasnosel'skiĭ [15] contain work instrumental to [84, Chaps. 7–8]. See [103] for more on the work of Krasnosel'skiĭ and [163] for an historical account by Zabreĭko. Secondly, Conley and coworkers developed their generalization of Morse- and Lyapunov theory in the late 1970s [35, 36], a topic we will only briefly touch on as it has been covered before.

What these works have in common is that they look for (algebraic) topological invariants that capture certain qualitative properties of spaces, maps, dynamical systems, and so forth. This viewpoint is at the core of this work.

---

[6] Although Rashevskii published similar work slightly earlier [128].

Summarizing, the study into dimension and equivalences resulted in the development of a host of topological tools. Building on these tools and in part due to the inherent difficulty of solving differential equations brought about the qualitative theory of dynamical systems and control.

This brief historical overview leaves us in the 1980s. The upcoming chapters will discuss how the control-, topology- and dynamical systems communities responded over the last 40 years and what can be learned from that body of work.

1.3   Case Study: Optimal Control on Lie Groups

To illustrate the developments we consider a problem simple enough to do explicit computations, but rich enough to be of importance. Specifically, we work with *Lie groups*, objects ubiquitous in engineering and physics [3, 14, 28, 109, 135].

A pair $(\mathsf{G}, \cdot)$, with G a set and $\cdot$ a binary operation, is a ***Lie group*** when

(i) the set G is a *group* under $\cdot$, that is, $g \cdot h \in \mathsf{G}$ for all $g, h \in \mathsf{G}$, there is an *identity element* $e \in \mathsf{G}$ such that $e \cdot g = g \cdot e = e$ for any $g \in \mathsf{G}$, for all $g \in \mathsf{G}$ there is an *inverse element* $g^{-1} \in \mathsf{G}$ such that $g \cdot g^{-1} = g^{-1} \cdot g = e$ and the $\cdot$ operation is associative;

(ii) the set G is a smooth manifold (informally, a set that is locally Euclidean and possesses a structure to make sense of derivations, see Chap. 2-3 for the details) and both multiplication and inversion are smooth maps.

When it clear from the context, the operator $\cdot$ is dropped, i.e., one writes gh instead of $g \cdot h$. For example, a Lie group of importance is the *special orthogonal group* $\mathsf{SO}(n, \mathbb{R}) = \{A \in \mathbb{R}^{n \times n} : A^{\mathsf{T}} A = I_n, \det(A) = 1\}$. Here the group operation $\cdot$ is matrix multiplication and for any $Q \in \mathsf{SO}(n, \mathbb{R})$, the corresponding inverse element is Q^{T} with the identity element being $e = I_n$, for I_n the identity matrix in $\mathbb{R}^{n \times n}$.

To every Lie group G corresponds a *Lie algebra*, denoted $\mathfrak{g}$, being a vector space identified with the tangent space of G at e (a vector space to be made precise in Chap. 3), denoted $\mathfrak{g} = T_e \mathsf{G}$. For example, $\mathfrak{so}(n, \mathbb{R}) = T_{I_n} \mathsf{SO}(n, \mathbb{R}) = \{X \in \mathbb{R}^{n \times n} : X^{\mathsf{T}} + X = 0\}$. Lie algebras are powerful for us in that they parametrize the tangent space of G at *any* $g \in \mathsf{G}$. To see this, pick any differentiable curve $t \mapsto \gamma(t) \in \mathsf{G}$ such that $\gamma(0) = e$. As we work with an abstract binary operator on G, it is convenient to define the ***left-translation*** map L_g by $h \mapsto L_g(h) = gh$ for any $g, h \in \mathsf{G}$. Now, define the curve $t \mapsto c(t) = L_g(\gamma(t)) \in \mathsf{G}$. Then, as $c(0) = g$, the derivative of c with respect to t satisfies $\dot{c}(t)|_{t=0} = DL_g(h)|_{h=e}\dot{\gamma}(t)|_{t=0} \in T_g\mathsf{G}$ such that we have the (tangent space) isomorphism $D(L_g)_e : T_e\mathsf{G} \to T_g\mathsf{G}$ for all $g \in \mathsf{G}$.

Now, let G be a compact connected Lie group and let $\langle \cdot, \cdot \rangle$ denote an Ad-invariant inner-product on $\mathfrak{g}$ (the adjective "Ad-*invariant*" can be ignored if unrecognized), which always exists as G is compact [82, Proposition 4.24]. A vector field X on G is said to be ***left-invariant*** when $D(L_g)_e X(e) = X(g) \in T_g\mathsf{G}$ for all $g \in \mathsf{G}$, differently put, the evaluation of the vector field at $e \in \mathsf{G}$, defines the vector field on all of G.

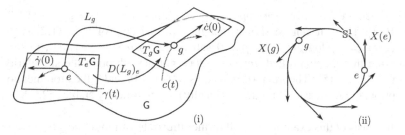

Fig. 1.3 Section 1.3 (i) curves and tangent spaces on the Lie group G; (ii) a vector field $X \in \text{Lie}(\mathbb{S}^1)$

The set of left-invariant vector fields is denoted by $\text{Lie}(G)$ and is isomorphic to $\mathfrak{g}$. For a visualization of the aforementioned concepts, see Fig. 1.3.

Consider for a set $\{X_0, \ldots, X_m\} \subset \text{Lie}(G)$ the input-affine control system on G

$$\frac{d}{dt} g(t) = X_0 (g(t)) + \sum_{i=1}^{m} X_i (g(t)) u_i, \tag{1.1a}$$

with $\text{span}\{X_1, \ldots, X_m\} = \text{Lie}(G)$ and the input vector $u \in \mathbb{R}^m$. As such, (1.1a) is *controllable* for controls $t \mapsto \mu(t) \in \mathbb{R}^m$ that are locally bounded and measurable [76, Theorem 7.1] (informally, one speaks of controllability of (1.1a) when for any $g_0, g_1 \in G$, there is a $T \geq 0$ and a map $\mu : [0, T] \to \mathbb{R}^m$ such that a solution $\varphi : [0, T] \times G \to G$ to (1.1a) under μ satisfies $\varphi(0, g_0) = g_0$ and $\varphi(T, g_0) = g_1$, for the precise definition see Chap. 5). Under these assumptions, one can study without loss of generality, the left-invariant control system

$$\frac{d}{dt} g(t) = D(L_{g(t)})_e(\mu(t)), \tag{1.1b}$$

where $\mu(t) = \sum_{i=1}^{m} E_i \mu_i(t)$ and $\text{span}\{E_1, \ldots, E_m\} = \mathfrak{g}$. Then, given a discount factor $\beta > 0$, and with abuse of notation the exponential function $e^{-\beta t}$, define the infinite-horizon optimal control problem

$$\begin{cases} \text{minimize} \quad \int_0^{\infty} e^{-\beta t} \left(d(e, g(t))^2 + \langle \mu(t), \mu(t) \rangle \right) dt \\ \quad \mu(\cdot) \\ \text{subject to} \quad (1.1b), \quad g(0) = g_0, \end{cases} \tag{1.2}$$

where the distance function $d(e, g)^2 = \langle \log(g), \log(g) \rangle$ between e and any $g \in G$ is defined with the aim of finding a feedback via (1.2) that stabilizes e in some sense. This construction is intended to generalize *Linear-quadratic regulation* (LQR) to nonlinear- systems and spaces, cf. [7, 8]. Note that $\log : G \to \mathfrak{g}$ is only well-defined over the subset of G where $\exp : \mathfrak{g} \to G$ is injective. Now, one can show that by construction of (1.2), one can appeal to *Hamilton–Jacobi–Bellman* (HJB) theory which provides necessary optimality conditions for (1.2), e.g., see [18, Theorem 10.2]. Then, it can be shown that the optimal controller in (1.2) is given by $\mu^\star(g(t)) = -p \log(g(t))$ for $p > 0$ satisfying $-\beta p - p^2 + 1 = 0$, under the assumption that the controlled trajectory does not pass through the singularity of $\exp : \mathfrak{g} \to G$, see [52, Theorem 4]. In fact, as $\mu^\star(e) = 0$, under the aforementioned assumption, the feedback $\mu^\star$ renders e (locally) asymptotically stable [52, Theorem 5].

On the basis of this example we will further illustrate various concepts, including, but not limited to: (i) the relation, or lack thereof, between controllability and the existence of continuous feedback; (ii) the source and (in)surmountability of discontinuous controllers; and (iii) the relation between the shape of the attractor and the domain of the dynamical system.

1.4   Content and Structure

This work surveys the inception, development and future of *topological obstructions* in the context of dynamical control systems. The aim is to present a unified and general treatment. As such, highly specialized results, as are known for surfaces, do not belong to the core of this work. Also, we largely focus on manifolds but indicate when results hold for more general topological spaces. Besides providing a review, a secondary goal of this work is to function as an invitation to the non-specialist.

In the past, a small number of reviews appeared, for example, on low dimensional systems by Dayawansa [39]. Close to us is the work by Sánchez-Gabites [133] and Sanjurjo [134], albeit mostly focused on shape and Conley theory. The work by Byrnes [30] and later by Kvalheim and Koditschek [88] also contain overviews, but mostly focused on generalizations of Brockett's condition. Topological obstructions are also briefly discussed in for example [141, Sect. 6], [19, Sect. 8], the monograph by Coron [37, Part 3], the monograph by Sontag [142, Sect. 5.9], the monograph by Zabczyk [162, Sect. 7.6] and the extensive survey by Vakhrameev [153] on the development of geometric methods in the study of controllability and optimal control. See also the introductions to [38, 108] and the voluminous work by Jonckheere [73] on algebraic topology and robust control. At last we highlight the thesis by Mayhew [104], containing hybrid- obstructions *and* solutions.

Regarding the exposition, we follow the philosophy as set forth in [51] and provide mostly arguments from differential topology with the aim of having an audience as large as possible that can follow and appreciate the complete development. Wherever insightful we do indicate how results can also be shown using arguments from algebraic topology. To further help the reader we provide ample examples, illustrations and references. Most of the results presented in this survey are already published and we systematically add a reference to which the reader can refer for more details. Some new results are added to complete those published and in this case we add a complete proof. Known proofs are occasionally presented to precisely show where assumptions are used and how to possibly relax them.

Although we will impose a smooth structure on our objects we stay in the topological realm and rarely assume knowledge of a metric on our spaces. The price to pay for this generality is that few things can be quantified.

We start by introducing a substantial amount of preliminary concepts from topology, dynamical systems and control theory. The benefit being that the core of this work can be described without technical clutter and in a somewhat self-contained manner. After presenting the topological obstructions, we also highlight how one might deal with these obstructions and what is considered future work. In particular, Chap. 2 introduces notions from general topology, e.g., homotopies and retraction theory whereas Chap. 3 introduces the prerequisite machinery for the Poincaré–Hopf theorem and the Bobylev–Krasnosel'skiǐ index theorem, that is, notions from differential topology like transversality, tubular neighbourhoods and index theory are discussed in detail. Then, Chap. 4 briefly presents material from algebraic topology and states how the Euler characteristic can be seen through different lenses, e.g., via self-intersections, combinatorially, or via homology theory. Chapter 5 introduces notions from dynamical systems theory like flows, vector fields and Lyapunov stability. Moreover, the dynamical control systems under consideration are defined. Chapter 6—the core of this work—is devoted to discussing topological obstructions to stability and stabilization. First, for equilibrium points, then for submanifolds and subsequently for generic sets. In particular, this section aims to show that just a few viewpoints allow for generalizing a wealth of results. Chapter 7 presents an overview of how to work with these obstructions, e.g., by allowing for singularities, time-dependent feedback or by employing techniques from hybrid control theory. Elaborating on some of the aforementioned tools, Chap. 8 offers a few generalizations and concludes with a list of future work.

Notation: We largely follow standard textbook notation, e.g., [51, 92, 142], but we state already that $p \in M^n$ denotes an element of a n-dimensional manifold M^n with the variable x being reserved for the state of a dynamical system. The symbols f and F are reserved to describe those dynamical systems, whereas g and G are used for general maps. When working with differential equations we use $d\xi(t)/dt$, $\dot{\xi}(t)$ or simply $\dot{\xi}$ to denote the *"time"*-derivative. Also, F_* will denote the *pushforward* of a map, whereas G_* denotes the *induced homomorphism* between groups. Any subtle difference in notation will always be accompanied by clarifying text.

References

1. Abraham R, Marsden JE (2008) Foundations of mechanics. American Mathematical Society, Providence
2. Andronov AA, Pontryagin L (1937) Systèmes grossieres. Dokl Akad Nauk SSSR 14(5):247–251
3. Arnold VI (1988) Mathematical methods of classical mechanics. Springer, Heidelberg
4. Barrow I, Haselden T (2015) Euclide's elements: the whole fifteen books compendiously demonstrated. FB & c Ltd, London

5. Barrow-Green J (1997) Poincaré and the three body problem. American Mathematical Society, Providence
6. Bellman R (1954) The theory of dynamic programming. B Am Math Soc 60(6):503–515
7. Bertsekas DP (2005) Dynamic programming and optimal control, vol 1. Athena Scientific, Belmont
8. Bertsekas DP (2007) Dynamic programming and optimal control, vol 2, 3rd edn. Athena Scientific, Belmont
9. Betti E (1871) Sopra gli spazi di un numero qualunque di dimensioni. Ann Mat Pur Appl 4(1):140–158
10. Bhatia NP, Szegö GP (1970) Stability theory of dynamical systems. Springer, Berlin
11. Birkhoff GD (1913) Proof of Poincaré's geometric theorem. T Am Math Soc 14–22
12. Birkhoff GD (1927) Dynamical systems. American Mathematical Society, Providence
13. Bissell C (2009) A history of automatic control. Springer, Berlin, pp 53–69
14. Bloch A (2015) Nonholonomic mechanics and control. Springer, New York
15. Bobylev N, Krasnosel'skiĭ M (1974) Deformation of a system into an asymptotically stable system. Autom Remote Control 35(7):1041–1044
16. Borsuk K (1931) Sur les rétractes. Fund Math 17:152–170
17. Bourbaki N (1989) General topology: chapters 1–4. Springer, Berlin
18. Bressan A (2011) Viscosity solutions of Hamilton-Jacobi equations and optimal control problems. Lecture notes
19. Brockett R (2014) The early days of geometric nonlinear control. Automatica 50(9):2203–2224
20. Brockett RW (1972) System theory on group manifolds and coset spaces. SIAM J Control 10(2):265–284
21. Brockett RW (1973) Lie algebras and Lie groups in control theory. In: Geometric methods in system theory. Springer, Dordrecht, pp 43–82
22. Brockett RW (1976) Some geometric questions in the theory of linear systems. IEEE T Automat Contr 21(4):449–455
23. Brockett RW (1983) Asymptotic stability and feedback stabilization. In: Differential geometric control theory. Birkhäuser, Boston, pp 181–191
24. Broer H (2004) KAM theory: the legacy of Kolmogorov's 1954 paper. B Am Math Soc 41(4):507–521
25. Brouwer LE (1911) Beweis der invarianz der dimensionenzahl. Math Ann 70(2):161–165
26. Brouwer LEJ (1911) Über abbildung von mannigfaltigkeiten. Math Ann 71(1):97–115
27. Brouwer LEJ (1912) Continuous one-one transformations of surfaces in themselves (5th communication). In: Proceedings of K Ned Akad B-Ph, vol 15, pp 352–361
28. Bullo F, Lewis AD (2004) Geometric control of mechanical systems. Springer, New York
29. Busoniu L, Babuska R, de Schutter B, Ernst D (2017) Reinforcement learning and dynamic programming using function approximators. CRC Press, Boca Raton
30. Byrnes CI (2008) On Brockett's necessary condition for stabilizability and the topology of Liapunov functions on $\mathbb{R}^n$. Commun Inf Syst 8(4):333–352
31. Cantor G (1932) Gesammelte Abhandlungen. Springer, Berlin
32. Casti JL (1982) Recent developments and future perspectives in nonlinear system theory. SIAM Rev 24(3):301–331
33. Chow W-L (1940) Über systeme von liearren partiellen differentialgleichungen erster ordnung. Math Ann 117(1):98–105
34. Ciesielski K (2012) The Poincaré-Bendixson theorem: from Poincaré to the XXIst century. Cent Eur J Math 10(6):2110–2128
35. Conley C, Zehnder E (1984) Morse-type index theory for flows and periodic solutions for Hamiltonian equations. Commun Pur Appl Math 37(2):207–253
36. Conley CC (1978) Isolated invariant sets and the Morse index. American Mathematical Society, Providence
37. Coron J-M (2007) Control and nonlinearity. American Mathematical Society, Providence

38. Coron J-M, Praly L, Teel A (1995) Feedback stabilization of nonlinear systems: sufficient conditions and Lyapunov and input-output techniques. Springer, London, pp 293–348
39. Dayawansa W (1993) Recent advances in the stabilization problem for low dimensional systems. In: Nonlinear control systems design 1992. Pergamon, Oxford, pp 1–8
40. Díaz JID, del Castillo LT (1999) A nonlinear parabolic problem on a Riemannian manifold without boundary arising in climatology. Collect Math 50(1):19–51
41. Dieudonné J (1978) Abrégé d'histoire des mathématiques: Fonctions elliptiques, analyse fonctionnelle, topologie, géométrie différentielle, probabilités, logique mathématique. Hermann, Paris
42. Dieudonné J (1989) A history of algebraic and differential topology, 1900–1960. Birkhäuser, Boston
43. Dinca G, Mawhin J (2021) Brouwer degree. Birkhäuser, Cham
44. Dirichlet GL (1946) Über die stabilität des gleichgewichts. J Reine Angew Math 32:85–88
45. Eilenberg S (1944) Singular homology theory. Ann Math 45(3):407–447
46. Fréchet MM (1906) Sur quelques points du calcul fonctionnel. Rend Circ Mat Palermo 22(1):1–74
47. Gamkrelidze RV (1999) Discovery of the maximum principle. J Dyn Control Syst 5(4):437–451
48. Gottlieb DH (1996) All the way with Gauss-Bonnet and the sociology of mathematics. Am Math Mon 103(6):457–469
49. Gouvêa FQ (2012) Was Cantor surprised? In: The best writing on mathematics 2012. Princeton University Press, Princeton, pp 216–233
50. Guckenheimer J, Holmes P (2013) Nonlinear oscillations, dynamical systems, and bifurcations of vector fields. Springer Science & Business Media, New York
51. Guillemin V, Pollack A (2010) Differential topology. American Mathematical Society, Providence
52. Gupta R, Kalabić UV, Bloch AM, Kolmanovsky IV (2018) Solution to the HJB equation for LQR-type problems on compact connected Lie groups. Automatica 95:525–528
53. Hadamard J (1897) Sur certaines propriétés des trajectoires en dynamique. J Math 5(3):331–387
54. Hadamard J (1898) Les surfaces à courbures opposées et leurs lignes géodesiques. J Math 5(3):27–73
55. Hartman P (1960) On local homeomorphisms of Euclidean spaces. Bol Soc Mat Mex 5(2):220–241
56. Hausdorff F (1914) Grundzüge der Mengenlehre. Veit, Leipzig
57. Haynes GW, Hermes H (1970) Nonlinear controllability via lie theory. SIAM J Control 8(4):450–460
58. Hermann R (1963) On the accessibility problem in control theory. In: International symposium on nonlinear differential equations and nonlinear mechanics. Academic Press, Cambridge, pp 325–332
59. Hermann R, Krener A (1977) Nonlinear controllability and observability. IEEE T Automat Contr 22(5):728–740
60. Hermann R, Martin CF (1977) Applications of algebraic geometry to systems theory—part I. IEEE T Automat Contr 22(1):19–25
61. Hilbert D (1930) Grundlagen der Geometrie. Teubner, Leipzig
62. Hirsch MW (1984) The dynamical systems approach to differential equations. B Am Math Soc 11(1):1–64
63. Hirsch MW, Pugh CC, Shub M (2006) Invariant manifolds. Springer, Berlin
64. Hirzebruch F (1999) Emmy Noether and topology
65. Holmes P (2005) Ninety plus thirty years of nonlinear dynamics: less is more and more is different. Int J Bifurcat Chaos 15(09):2703–2716
66. Hopf H (1925) Über die curvatura integra geschlossener hyperflächen. Math Ann 95(1):340–367
67. Hopf H (1926) Vektorfelder in n-dimensionalen-mannigfaltigkeiten. Math Ann 96:225–250

68. Hurewicz W (1935) Homotopie und homologiegruppen. P K Akad Wet-Amsterd. 38:521–528
69. Hurwitz A (1895) Über die bedingungen, unter welchen eine gleichung nur wurzeln mit negativen reellen theilen besitzt. Math Ann 46(2):273–284
70. Isidori A (1985) Nonlinear control systems: an introduction. Springer, Berlin
71. Isidori A (2013) The zero dynamics of a nonlinear system: from the origin to the latest progresses of a long successful story. Eur J Control 19(5):369–378
72. James IM (1999) History of topology. Elsevier, Amsterdam
73. Jonckheere EA (1997) Algebraic and differential topology of robust stability. Oxford University Press, Oxford
74. Jordan C (1866) Sur la déformation des surfaces. J Math Pure Appl 105–109
75. Jurdjevic V, Quinn JP (1978) Controllability and stability. J Differ Equ 28(3):381–389
76. Jurdjevic V, Sussmann HJ (1972) Control systems on lie groups. J Differ Equ 12(2):313–329
77. Jury EI (1996) Remembering four stability theory pioneers of the nineteenth century. IEEE T Automat Contr 41(9):1242
78. Kalman R (1960) Contributions to the theory of optimal control. Bol Soc Mat Mex 102–119
79. Kalman RE, Bucy RS (1961) New results in linear filtering and prediction theory. J Basic Eng-T ASME 83(1):95–108
80. Kelley A (1967) The stable, center-stable, center, center-unstable, unstable manifolds. J Differ Equ 3(4):546–570
81. Khalil HK (2002) Nonlinear systems. Prentice Hall
82. Knapp AW (2002) Lie groups beyond an introduction. Birkhäuser, Boston
83. Koditschek DE (2021) What is robotics? Why do we need it and how can we get it? Annu Rev Control 4:1–33
84. Krasnosel'skiĭ A, Zabreiko PP (1984) Geometrical methods of nonlinear analysis. Springer, Berlin
85. Krasnosel'skiĭ MA (1968) The operator of translation along the trajectories of differential equations. American Mathematical Society, Providence
86. Krasovskiĭ N (1963) Stability of motion. Stanford University Press, Palo Alto
87. Kurzweil J (1963) On the inversion of Ljapunov's second theorem on stability of motion. AMS Transl Ser 2(24):19–77
88. Kvalheim MD, Koditschek DE (2022) Necessary conditions for feedback stabilization and safety. J Geom Mech
89. La Salle J, Lefschetz S (1961) Stability by Liapunov's direct method with applications. Academic Press, New York
90. Lagrange JL (1788) Mécanique Analytique. La Veuve Desaint, Paris
91. LaSalle J (1960) Some extensions of Liapunov's second method. IRE T Circuit Theor 7(4):520–527
92. Lee JM (2011) Introduction to topological manifolds. Springer, New York
93. Lefschetz S (1926) Intersections and transformations of complexes and manifolds. T Am Math Soc 28(1):1–49
94. Lefschetz S (1937) On the fixed point formula. Ann Math 819–822
95. Lefschetz S (1957) Differential equations: geometric theory. Interscience Publishers, New York
96. Leine RI (2010) The historical development of classical stability concepts: Lagrange. Poisson and Lyapunov stability. Nonlinear Dyn 59(1):173–182
97. Levi M (1988) Stability of the inverted pendulum'a topological explanation. SIAM Rev 30(4):639–644
98. Lewis AD (2001) A brief on controllability of nonlinear systems
99. Liapunov A (1892) A general task about the stability of motion. Dissertation, University of Kharkov
100. Lobry C (1970) Contrôlabilité des systèmes non linéaires. SIAM J Control 8(4):573–605
101. Lobry C (1974) Controllability of nonlinear systems on compact manifolds. SIAM J Control 12(1):1–4

102. Lyapunov A (1992) The general problem of the stability of motion. Int J Control 55(3):531–773
103. Mawhin J (2000) Mark A. Krasnosel'skii and nonlinear analysis: a fruitful love story. In: Coll Mark Aleksandrovich Krasnose'skii. To the 80th anniversary of his birth. Digest of articles, IITP RAS, pp 80–97
104. Mayhew CG (2010) Hybrid control for topologically constrained systems. PhD thesis, University of California, Santa Barbara
105. McLennan A (2018) Advanced fixed point theory for economics. Springer, Singapore
106. Möbius AF (1863) Theorie der elementaren verwandtschaft. Berichte über die Verhandlungen der Königlich Sächsischen Gesellschaft der Wissenschaften, Mathematisch-physikalische Klasse 15:19–57
107. Morse M (1925) Relations between the critical points of a real function of n independent variables. T Am Math Soc 27(3):345–396
108. Moulay E, Hui Q (2011) Conley index condition for asymptotic stability. Nonlinear Anal-Theor 74(13):4503–4510
109. Murray RM, Li Z, Sastry SS (1994) A mathematical introduction to robotic manipulation. CRC Press, Boca Raton
110. Nijmeijer H, van der Schaft A (1990) Nonlinear dynamical control systems. Springer, New York
111. Noether E (1926) Ableitung der elementarteilertheorie aus der gruppentheorie, nachrichten der 27 januar 1925. Jahresbericht der Deutschen Mathematiker-Vereinigung 34:104
112. Noether E, Cavailles J (1937) Briefwechsel Cantor-Dedekind. Hermann, Paris
113. Orsi R, Praly L, Mareels I (2003) Necessary conditions for stability and attractivity of continuous systems. Int J Control 76(11).1070–1077
114. Outerelo E, Ruiz JM (2009) Mapping degree theory. American Mathematical Society, Prov idence
115. Peano G (1890) Sur une courbe, qui remplit toute une aire plane. Math Ann 36(1):157–160
116. Peixoto MM (1962) Structural stability on two-dimensional manifolds. Topology 1(2):101–120
117. Pliss VA (1964) A reduction principle in the theory of stability of motion. Izv Akad Nauk SSSR Ser Mat 28(6):1297–1324
118. Poincaré H (1890) Sur le problème des trois corps et les équations de la dynamique. Acta Math 13(1):1–270
119. Poincaré H (1892) Les Méthodes Nouvelles de la Mécanique Céleste, vol 1. Gauthier-Villars, Paris
120. Poincaré H (1893) Les Méthodes Nouvelles de la Mécanique Céleste, vol 2. Gauthier-Villars, Paris
121. Poincaré H (1895) Analysis situs. J Ec Polytech-Math 1:1–121
122. Poincaré H (1899) Les Méthodes Nouvelles de la Mécanique Céleste, vol 3. Gauthier-Villars, Paris
123. Poincaré H (1910) The future of mathematics. Monist 20(1):76–92
124. Poincaré H (2010) Papers on topology: analysis situs and its five supplements. American Mathematical Society, Providence
125. Poincaré H (2017) The three-body problem and the equations of dynamics: Poincar's foundational work on dynamical systems theory. Springer, Cham
126. Polekhin I (2018) On topological obstructions to global stabilization of an inverted pendulum. Syst Control Lett 113:31–35
127. Pontryagin L, Boltyansky V, Gamkrelidze R, Mishchenko E (1962) The mathematical theory of optimal processes. Wiley, New York
128. Rashevskii P (1938) Joinability of any two points of a completely nonholonomic space by an admissible line. Uch Zapiski Ped Inst Libknexta Ser Fiz-Mat (2), 83:94
129. Riemann B (1851) *Grundlagen für eine allgemeine Theorie der Functionen einer veränderlichen complexen Grösse*. Inauguraldissertation, Göttingen
130. Riemann B (1892) Gesammelte Mathematische Werke. Teubner, Leipzig

131. Riesz F (1908) Stetigkeitsbegriff und abstrakte mengenlehre. Atti del IV congresso inter-
     nazionale dei matematici, Bologna 2:18–24
132. Ruelle D, Takens F (1971) On the nature of turbulence. Commun Math Phys 20:167–192
133. Sánchez-Gabites J (2008) Dynamical systems and shapes. RACSAM REV R Acad A
     102(1):127–159
134. Sanjurjo JMR (2008) Shape and Conley index of attractors and isolated invariant sets. In:
     Differential equations, chaos and variational problems. Birkhäuser, Basel, pp 393–406
135. Sastry S (1999) Nonlinear systems. Springer, New York
136. Schoukens J, Ljung L (2019) Nonlinear system identification: a user-oriented road map. IEEE
     Contr Syst Mag 39(6):28–99
137. Siegberg HW (1981) Some historical remarks concerning degree theory. Am Math Mon
     88(2):125–139
138. Smale S (1961) Generalized Poincare's conjecture in dimensions greater than four. Ann Math
     74(2):391–406
139. Smale S (1967) Differentiable dynamical systems. B Am Math Soc 73(6):747–817
140. Sontag E, Sussmann H (1980) Remarks on continuous feedback. In: Proceedings of IEEE
     conference on decision control, pp 916–921
141. Sontag ED (1990) Feedback stabilization of nonlinear systems. Birkhäuser, Boston, pp 61–81
142. Sontag ED (1998) Mathematical control theory: deterministic finite dimensional systems.
     Springer, New York
143. Steen LA, Seebach JA (1978) Counterexamples in topology. Springer, New York
144. Stillwell J (2012) Poincaré and the early history of 3-manifolds. B Am Math Soc 49(4):555–
     576
145. Strogatz SH (2014) Nonlinear dynamics and chaos. Westview Press, Boulder
146. Sussmann HJ (1979) Subanalytic sets and feedback control. J Differ Equ 31(1):31–52
147. Sussmann HJ, Jurdjevic V (1972) Controllability of nonlinear systems. J Differ Equ 12(1):95–
     116
148. Tabak J (2011) Beyond geometry: a new mathematics of space and form. Facts on File, New
     York
149. Thom R (1954) Quelques propriétés globales des variétés différentiables. Comment Math
     Helv 28(1):17–86
150. Thom R (2018) Structural stability and morphogenesis. CRC Press, Boca Raton
151. Thorne KS, Misner CW, Wheeler JA (1973) Gravitation. Freeman, Reading
152. Urysohn P (1925) Mémoir on cantorian multiplicities. Fund Math 1(7):30–137
153. Vakhrameev S (1995) Geometrical and topological methods in optimal control theory. J Math
     Sci 76(5):2555–2719
154. Veblen O, Whitehead JHC (1932) The foundations of differential geometry. Cambridge Uni-
     versity Press, Cambridge
155. Verhulst F (2012) Henri Poincaré: impatient genius. Springer Science & Business Media,
     New York
156. Westervelt ER, Grizzle JW, Chevallereau C, Choi JH, Morris B (2018) Feedback control of
     dynamic bipedal robot locomotion. CRC Press, Boca Raton
157. Whitney H (1936) Differentiable manifolds. Ann Math 37(3):645–680
158. Whitney H (1937) Topological properties of differentiable manifolds. B Am Math Soc
     43(12):785–805
159. Wilson FW (1969) Smoothing derivatives of functions and applications. T Am Math Soc
     139:413–428
160. Wilson FW Jr (1967) The structure of the level surfaces of a Lyapunov function. J Differ Equ
     3(3):323–329
161. Wonham WM (1979) Linear multivariable control. Springer, New York
162. Zabczyk J (2020) Mathematical control theory. Birkhäuser, Cham
163. Zabrejko P (1997) Rotation of vector fields: definition, basic properties, and calculation. In:
     Topological nonlinear analysis II. Birkhäuser, Boston, pp 445–601

Chapter 2
General Topology

2.1 Topological Spaces

If a set X and a collection of its subsets τ satisfy the following three properties (i) τ contains X and the empty set $\emptyset$; (ii) τ is closed under finite intersections; (iii) τ is closed under arbitrary unions; then, the pair (X, τ) is called a *topological space*. The elements of τ are said to be *open* and their complements in X are said to be *closed*. We assume all our topological spaces (X, τ) to be *Hausdorff*, that is, for any two points $p_1, p_2 \in X$ there exist open neighbourhoods $U_1, U_2 \in \tau$ of p_1 and p_2, respectively, such that $U_1 \cap U_2 = \emptyset$. In what follows we frequently drop the explicit declaration of the topology τ. A point $p \in X$ is called a *limit* of the sequence $\{p_k\}_{k \geq 0}$ if for any open neighbourhood U of p there is a $K \in \mathbb{N}$ such that $p_k \in U$ for all $k \geq K$. As X is a Hausdorff space, this limit is unique, which is important in the context of defining dynamical systems and their stability. Moreover, as we will appeal to Whitney's approximation theorems, we assume all our topological spaces (X, τ) to be *second countable*, that is, there is a set $\mathscr{B} \subseteq \tau$ such that every element in τ can be written as a union of countably many elements in $\mathscr{B}$, i.e., τ admits a countable basis. Then, we call the topological space (X, τ) a n-dimensional *topological manifold*, when for each $p \in X$ there is an open neighbourhood $U \in \tau$ of p such that U is *homeomorphic* to $\mathbb{R}^n$ (or equivalently, some open set of $\mathbb{R}^n$), that is, there is a continuous bijection between U and $\mathbb{R}^n$ with the inverse of this map also being continuous (see below). When these homeomorphisms fail to exist, but *do* exist when elements of τ are also allowed to be homeomorphic to open subsets of $\mathbb{H}^n = \{p \in \mathbb{R}^n : p_n \geq 0\}$, X is said to be a manifold with *boundary*, frequently denoted as $\partial X \neq \emptyset$. Indeed, $\partial(\partial X) = \emptyset$.

Example 2.1 (*The standard topology on* $\mathbb{R}^n$) Let $\| \cdot \|$ be a norm on $\mathbb{R}^n$ and let $\mathbb{B}_r^n(p) = \{y \in \mathbb{R}^n : \|p - y\| < r\}$ be an open ball in $\mathbb{R}^n$. The collection of all these open balls gives rise to a topology on $\mathbb{R}^n$, called the norm topology, or the *standard topology*, denoted τ_{std}. Now it can be shown that the set of all open balls $\mathbb{B}_r^n(p)$, with a rational radius r, centred at a point p with rational coordinates, is a countable basis for the standard topology [4, Chap. IV]. As any two points $p_1, p_2 \in \mathbb{R}^n$ admit

© The Author(s) 2023

W. Jongeneel and E. Moulay, *Topological Obstructions to Stability and Stabilization*,
SpringerBriefs in Control, Automation and Robotics,
https://doi.org/10.1007/978-3-031-30133-9_2

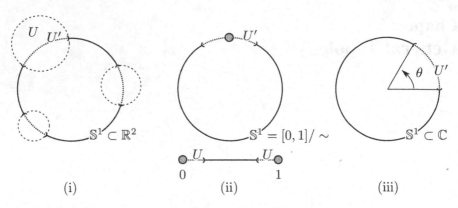

Fig. 2.1 Example 2.2, for the subspace- (i), quotient- (ii) and the standard topology (iii) on $\mathbb{S}^1$, we show a typical open set U'. When applicable, U denotes the corresponding open set in the topological space the topology on $\mathbb{S}^1$ is inherited from

open non-intersecting neighbourhoods $\mathbb{B}_r^n(p_1)$, $\mathbb{B}_r^n(p_2) \in \tau_{\text{std}}$ for $r = \frac{1}{2}\|p_1 - p_2\|$, it readily follows that $(\mathbb{R}^n, \tau_{\text{std}})$ is Hausdorff and second-countable. As any open ball is homeomorphic to $\mathbb{R}^n$, e.g., consider without loss of generality $\mathbb{B}_1^n(0)$ and see that the homeomorphism $\varphi : \mathbb{B}_1^n(0) \to \mathbb{R}^n$ is given by $\varphi : p \mapsto p/(1 - \|p\|)$ with the inverse map $\varphi^{-1} : y \mapsto y/(1 + \|y\|)$, it follows that $(\mathbb{R}^n, \tau_{\text{std}})$ is in fact a topological manifold.

Example 2.2 (*Topologies on the circle* $\mathbb{S}^1$) When looking at the circle as a subset of the plane, i.e., $\mathbb{S}^1 = \{x \in \mathbb{R}^2 : \|x\|_2 = 1\}$, one can define a topology on $\mathbb{S}^1$ via a topology on $\mathbb{R}^2$. Generally, let (X, τ) be a topological space and let $A \subseteq X$, then $\tau_A = \{A \cap U : U \in \tau\}$ is the **subspace topology** on A. The circle can also be described as $\mathbb{S}^1 = \mathbb{R}/\mathbb{Z}$ or $\mathbb{S}^1 = [0, 1]/\sim$ for $0 \sim 1$, that is, one identifies all integers. Now again, the topology on $\mathbb{R}$ can be used to generate a topology on $\mathbb{R}/\mathbb{Z}$. Generally, let $\sim$ be an equivalence relation on the topological space (X, τ) and define the surjective map $q : X \to X/\sim$, then, the **quotient topology** on $X/\sim$ is defined as $\tau_{/\sim} = \{U \subseteq X/\sim : q^{-1}(U) \in \tau\}$. A third option would be to directly employ open sets of the form $\{e^{i\theta} : \theta \in (a, b) \subseteq [0, 2\pi]\} \subset \mathbb{C}$ and proceed as in Example 2.1. See Fig. 2.1 for a visualization of these topologies.

A function $f : \mathbb{R} \to \mathbb{R}$ is said to be *continuous* at $x \in \mathbb{R}$ when for each $\varepsilon > 0$ there is $\delta > 0$ such that for all $y \in \mathbb{R}$ satisfying $|x - y| < \delta$ one has $|f(x) - f(y)| < \varepsilon$. Some refer to this construction as the $\varepsilon - \delta$ definition of continuity. Imposing a topology on spaces X and Y allows for generalizing the notion of continuity beyond Euclidean spaces. Let (X, τ) and (Y, τ') be topological spaces, then $f : X \to Y$ is said to be **continuous** when for each $V \in \tau'$ the preimage under f is a contained in τ, i.e., $f^{-1}(V) = \{p \in X : f(p) \in V\} \in \tau$. Indeed, under the standard topology on $\mathbb{R}$, one recovers the $\varepsilon - \delta$ definition. Another concept of importance is that of *compactness*. An *open cover* of a topological space (X, τ) is a collection of open sets $\mathscr{U} = \{U_j\}_{j \in \mathcal{J}}$ with $U_j \subseteq X$ for all $j \in \mathcal{J}$, such that $X = \cup_{j \in \mathcal{J}} U_j$. Then, if a subset of $\mathscr{U}$ still covers X, this subset is said to be a *subcover*. Now a topological space X is **compact** when every open cover of X has a *finite* subcover. The notion of compactness is fundamental in topology since for any continuous map $f : X \to Y$ between topological spaces X and Y, when X is compact, so is $f(X)$ [9, Theorem 4.32]. A useful result is the Heine-Borel theorem, stating that a subset of $\mathbb{R}^n$ is compact if and only if it is closed and bounded [9, Theorem 4.40]. One should observe that continuity and compactness can be in conflict, i.e., a *fine* topology is desirable from a continuity point of view, yet a *coarse* topology is easier to work with when it comes to compactness.

Regarding notation, we will drop the explicit dependency on τ as the upcoming material is invariant under the particular choice of the topology, as long as the topology satisfies the properties as highlighted above. Besides, the dimension (of the component(s) under consideration) is frequently added by means of a superscript, i.e., X^n denotes a n-dimensional topological manifold. Unless stated otherwise, n will be finite. As mentioned above, maps of interest are *homeomorphisms*, i.e., continuous bijections with a continuous inverse. When two objects are homeomorphic, we speak of **topological equivalence**, denoted $\simeq_t$. Here, mapping the interval $[0, 1)$ to the circle $\mathbb{S}^1$ is the prototypical example of a map that is continuous and one-to-one, yet not a homeomorphism as the inverse cannot be chosen to be continuous.

2.2 Homotopy and Retractions

It turns out that many topological invariants (under homeomorphims) are invariant under a weaker notion; that of *homotopy*.[1] Let X and Y be topological spaces with g_1 and g_2 continuous maps from X to Y. A continuous map $H : [0, 1] \times X \to Y$ is said to be a **homotopy** from g_1 to g_2 when for all $p \in X$ we have $H(0, p) = g_1(p)$ and $H(1, p) = g_2(p)$. If such a map exists, g_1 and g_2 are homotopic, which is an *equivalence relation*, denoted $g_1 \simeq_h g_2$. Moreover, if H is stationary with respect to some set $A \subseteq X$, that is, $H(t, p) = g_1(p) = g_2(p)$ for all $p \in A$ and $t \in [0, 1]$, then, H is a **homotopy relative to** A. We note that not only homotopies give rise to an equivalence class, but also homotopies relative to some subset [14, p. 24]. Two topological spaces X and Y are called **homotopy equivalent**, or simply *homotopic*, when there are continuous maps $g_1 : X \to Y$, $g_2 : Y \to X$ such that $g_1 \circ g_2 \simeq_h \mathrm{id}_Y$ and $g_1 \circ g_2 \simeq_h \mathrm{id}_X$, e.g., generalizing the concept of a homeomorphism to maps that are not necessarily invertible. It is imperative to remark that when colloquially referring to "*the topology of a space* X" one commonly refers to the homotopy type of X.

Definition 2.1 (*Retractions*) Given a topological space X, a subset $A \subseteq X$ is a **retract** of X if there is a continuous map $r : X \to A$, called a retraction, such that $r \circ \iota_A = \mathrm{id}_A$, for $\iota_A : A \hookrightarrow X$ the inclusion map. The retraction is said to be a **deformation retract** when $\iota_A \circ r \simeq_h \mathrm{id}_X$. We speak of a **strong deformation retract** when the homotopy is relative to A. On the other hand, A is **weak deformation retract** of X if every open neighbourhood $U \subseteq X$ of A contains a strong deformation retract V of X such that $A \subseteq V$.

---

[1] One can argue that homotopy theory is a field of its own and not merely a branch of topology [1].

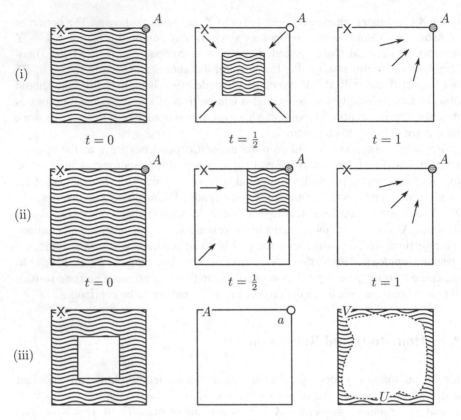

(i)

$t = 0$ $t = \frac{1}{2}$ $t = 1$

(ii)

$t = 0$ $t = \frac{1}{2}$ $t = 1$

(iii)

Fig. 2.2 Definition 2.1, (i) a deformation retract of X onto the point A; (ii) a strong deformation retract of X onto the point A; (iii) a weak deformation retract of X onto the set A, with A being the boundary of the cube without the point a, U an open neighbourhood of A and $V \subset U$ a strong deformation retract of X

A deformation retraction maps all of X, continuously, to A, but with A free to move throughout the process. On the other hand, a *strong* deformation retract keeps A stationary, see also Fig. 2.2. A mere retraction to a point is not particularly interesting as one can retract to any point via the constant map. As will be clarified below, deformation retracts, however, relate to stability notions indeed. For more on retraction theory, see [2, 7], it is imperative to remark that the literature does not agree on the terminology used in Definition 2.1 cf. [5].

Lemma 2.1 (Subset deformation retract) *Let both A and B be deformation retracts of C. Then, if $A \subseteq B$, A is a deformation retract of B.*

Proof As C deformation retracts on $A \subseteq C$ there is a map $r_A : C \to A$ such that $r_A \circ \iota_{AC} = \mathrm{id}_A$, $\iota_{AC} \circ r_A \simeq_h \mathrm{id}_C$ for $\iota_{AC} : A \hookrightarrow C$. Similarly for $B \subseteq C$, there is a map $r_B : C \to B$ such that $r_B \circ \iota_{BC} = \mathrm{id}_B$, $\iota_{BC} \circ r_B \simeq_h \mathrm{id}_C$. Now construct the map $r : B \to A$ via the inclusion map $\iota_{BC} : B \hookrightarrow C$, that is, $r = r_A \circ \iota_{BC}$. As $A \subseteq B$, we have that $r \circ \iota_{AB} = r_A \circ \iota_{BC} \circ \iota_{AB} = r_A \circ \iota_{AC} = \mathrm{id}_A$. Moreover, as $\iota_{BC} \circ \iota_{AB} \circ r_A \simeq_h \mathrm{id}_C$ and $r_B \circ \iota_{BC} = \mathrm{id}_B$ we have that $\iota_{AB} \circ r \simeq_h \mathrm{id}_B$, as desired.

For more on the relation between homotopies and deformation retractions, see, [7], [14, Chap. 1], [5, Chap. 0] or [10, Chap. 7].

When for a closed subset $A \subseteq X$ there is an open neighbourhood $U \subseteq X$ of A such that A is any retraction type from Definition 2.1 of U, then A is said to be a **neighbourhood retract**, of that particular type, e.g., a neighbourhood deformation retract, reconsider Fig. 2.2(iii).

Lemma 2.2 (Neighbourhood retracts [12, Theorem 4]) *Let $A \subseteq X$ be a weak deformation retract of $B \subseteq X$, then the following hold:*

- *(i) if A is a neighbourhood retract of X, then A is a retract of B;*
- *(ii) if A is a neighbourhood deformation retract of X, then A is a deformation retract of B;*
- *(iii) if A is a strong neighbourhood deformation retract of X, then A is a strong deformation retract of B.*

The intuition behind Lemma 2.2 is that B strongly deformation retracts onto a *neighbourhood* of A, which can be subsequently retracted to A itself.

The prototypical retraction example is that of the sphere $\mathbb{S}^{n-1}$ being a strong deformation retract of the punctured Euclidean space $\mathbb{R}^n \setminus \{0\}$. To see this, consider $r(p) = p / \|p\|_2$ and let the homotopy, relative to $\mathbb{S}^{n-1}$, be the convex combination of r and $\mathrm{id}_{\mathbb{R}^n}$, that is, $H(t, p) = tr(p) + (1 - t)p$. See Example 3.1 for a retraction in the context of vector bundles, Example 5.3 for a homotopy in the context of Lyapunov functions and Example 6.10 for strong deformation retracts of Lie groups.

A set S is **contractible** when id_S is homotopic to a constant map. Equivalently, S is homotopy equivalent to a point or a point $p \in S$ is a deformation retract of S. For example, X in Fig. 2.2(i) is contractible, while X in Fig. 2.2(iii) is not. Note, contractability does not imply that the deformation is *strong* [5, Exercise 0.6].

Remark 2.1 (*On contractible sets*) One might expect that all n-dimensional contractible sets are homeomorphic to $\mathbb{R}^n$. In 1935, Whitehead provided the first counterexample. Namely, there is an open, 3-dimensional manifold which is contractible but not homeomorphic to $\mathbb{R}^3$, see [15]. Although we focus on the finite-dimensional setting, more counter-intuitive phenomena appear in the infinite-dimensional setting. For example, $\mathbb{S}^\infty$ is contractible [5, Example 1B.3].

2.3 Comments on Triangulation

Motivated by Morse [3, p. 913], triangulations were formally introduced by Cairns, with further initial work by Brouwer, Freudenthal and Whitehead [8, Chap. 15]. A topological space X is called **triangulable** when the space is homeomorphic to some

Fig. 2.3 Adding the lines
ℓ_1, ℓ_2 and ℓ_3 preserves the
Euler characteristic

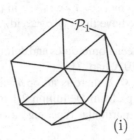
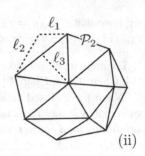

(i)   (ii)

polyhedron $\mathcal{P}$. Then, the *Euler characteristic* for surfaces of polyhedra is given by
$\chi(\mathcal{P}) = \mathcal{V} - \mathcal{E} + \mathcal{F}$, for $\mathcal{V}$ the number of vertices (0-dimensional), $\mathcal{E}$ the number of
edges (1-dimensional) and $\mathcal{F}$ the number of faces[2] (2-dimensional) of the polyhedron
$\mathcal{P}$ at hand. It turns out that this number $\chi(\mathcal{P})$ equals $2 - 2g$, for g the number of
holes in $\mathcal{P}$ and is independent of how one selects the triangulation, as such, χ is a
topological invariant of X, see Fig. 2.3. This invariance is why in what follows one
will keep seeing alternating sums akin to $\chi(\mathcal{P})$. Studying a topological space X via a
naïve triangulation, however, requires attention above dimension 3, those topological
spaces do not have a canonical triangulation e.g., see [13], [9, Chap. 5] for more on
the so-called *Hauptvermutung*.

For further references on general topology, see [5, 7, 9, 14] and see [11] for how
homotopies appeared in the context of robust control, albeit not explicitly.

References

1. Barwick C (2017) The future of homotopy theory
2. Borsuk K (1967) Theory of retracts. Państwowe Wydawn. Naukowe, Warszawa
3. Bott R (1980) Marston Morse and his mathematical works. B Am Math Soc 3(3):907–950
4. Bourbaki N (1989) General topology: chapters 1–4. Springer, Berlin
5. Hatcher A (2002) Algebraic topology. Cambridge University Press, Cambridge
6. Hilton P, Pedersen J (1994) Euler's theorem for polyhedra: a topologist and geometer respond
    (with a response from Grünbaum and Shephard). Am Math Mon 101:959–962
7. Hu ST (1965) Theory of retracts. Wayne State University Press, Detroit
8. James IM (1999) History of topology. Elsevier, Amsterdam
9. Lee JM (2011) Introduction to topological manifolds. Springer, New York
10. Lee JM (2012) Introduction to smooth manifolds. Springer, New York
11. Megretski A, Rantzer A (1997) System analysis via integral quadratic constraints. IEEE T
    Automat Contr 42(6):819–830
12. Moulay E, Bhat SP (2010) Topological properties of asymptotically stable sets. Nonlinear
    Anal-Theor 73(4):1093–1097
13. Ranicki AA (ed) The Hauptvermutung book. In: Casson AJ, Sullivan DP, Armstrong MA,
    Rourke CP, Cooke GE. Springer, Dordrecht
14. Spanier EH (1981) Algebraic topology. McGraw-Hill, New York
15. Whitehead J (1935) A certain open manifold whose group is unity. Q J Math os-6(1):268–279

---

[2] See [6] for comments on prevailing misunderstandings of $\chi(\mathcal{P})$.

Chapter 3
Differential Topology

3.1 Differentiable Structures

To make sense of differentiation on a topological manifold, we need to provide additional structure. A pair (U, φ) with U open in M^m and φ a homeomorphism from U to some open subset of $\mathbb{R}^m$ is called a **chart**. Then, for any $p \in U$, $\varphi(p) = (x^1(p), \dots, x^m(p)) \in \mathbb{R}^m$ are said to be **local coordinates** of p on U, with the inverse map sometimes called a *parametrization*. A pair of charts (U_1, φ_1) and (U_2, φ_2) is C^r-compatible when either $U_1 \cap U_2 = \emptyset$ or $\varphi_2 \circ \varphi_1^{-1} : \varphi_1(U_1 \cap U_2) \to \varphi_2(U_1 \cap U_2)$ is C^r smooth. A collection of charts $\{(U_i, \varphi_i)\}_{i \in \mathcal{I}}$ such that $\mathsf{M} = \cup_{i \in \mathcal{I}} U_i$ and all charts are C^r-compatible is called a C^r-**smooth atlas**. Now we can define a C^r-smooth **maximal atlas**, denoted $\bar{\mathscr{A}}$, as the atlas that contains all charts C^r-compatible with the elements of $\mathscr{A}$. Then, we say that M is a C^r-smooth manifold, or simply a C^r manifold, when M admits a C^r-smooth maximal atlas $\bar{\mathscr{A}} = \{(U_i, \varphi_i)\}_{i \in \mathcal{I}}$ for some $r \in \mathbb{N} \cup \{\infty\} \cup \{\omega\}$. As such, one can call a topological manifold a C^0 manifold. See [27, Example 1.4, Example 1.31] for the construction of a smooth structure on $\mathbb{S}^{n-1}$. It is imperative, however, to point out that one rarely constructs atlases explicitly, their mere existence usually suffices. We speak of *smooth* manifolds when $r \geq 1$. There is no need to further classify these spaces as for $r \geq 1$, every C^r manifold is C^r diffeomorphic to a C^∞ manifold [18, Theorem 2.2.10].

Given a smooth manifold M^m, then $T_p\mathsf{M}^m$ denotes the **tangent space** of M^m at the point $p \in \mathsf{M}^m$, that is, $T_p\mathsf{M}^m = \{\dot{\gamma}(t)|_{t=0} : t \mapsto \gamma(t) \in \mathsf{M}$ is a curve differentiable at 0 with $\gamma(0) = p\}$. Now by considering equivalence classes of curves, with respect to $\dot{\gamma}(t)_{t=0}$ in coordinates, one can show that $T_p\mathsf{M}^m$ has a m-dimensional vector space structure [27, Chap. 3]. The disjoint union $T\mathsf{M}^m = \sqcup_{p \in \mathsf{M}^m} T_p\mathsf{M}^m$ is the **tangent bundle** of M^m and is a smooth $2m$-dimensional manifold itself [27, Proposition 3.18].

3.2 Submanifolds and Transversality

Given two smooth manifolds M^m and N^n with $m \geq n$, let $G : \mathsf{M}^m \to \mathsf{N}^n$ be a smooth map, then, $q = G(p) \in \mathsf{N}^n$ is a **regular value** if the differential of G at p, $DG_p : T_p\mathsf{M}^m \to T_{G(p)}\mathsf{N}^n$, is surjective for all p such that $G(p) = q$. The points $p \in \mathsf{M}^m$

© The Author(s) 2023

W. Jongeneel and E. Moulay, *Topological Obstructions to Stability and Stabilization*,
SpringerBriefs in Control, Automation and Robotics,
https://doi.org/10.1007/978-3-031-30133-9_3

where this surjectivity condition fails are called ***critical points*** of G on M^m. Now it follows from Sard's theorem that regular values are *generic*[1] [27, Theorem 6.10]. Similarly, one can define regular points and critical values. The critical points of a smooth function $g : \mathsf{M} \to \mathbb{R}$ are all points $p \in \mathsf{M}$ such that $Dg_p = 0$.

Again, let M and N be smooth manifolds and let $G : \mathsf{M} \to \mathsf{N}$ be a smooth map. The map G is called a smooth ***submersion*** when DG_p is surjective for all $p \in \mathsf{M}$. Similarly, G is a smooth ***immersion*** when DG_p is injective for all $p \in \mathsf{M}$. The map G is called a smooth ***embedding*** when G is an immersion and M is homeomorphic to its image $G(\mathsf{M})$. Let the subset $\mathsf{S} \subseteq \mathsf{M}$ be a manifold under the subspace topology, then, S is said to be an ***embedded submanifold*** when the inclusion $\iota_\mathsf{S} : \mathsf{S} \hookrightarrow \mathsf{M}$ is a smooth embedding. When irrelevant or unknown, the adjective "*embedded*" is omitted, the same is true for the declaration of a particular map ι_S, the mere existence of some embedding usually suffices, i.e., we simply write $\mathsf{S} \hookrightarrow \mathsf{M}$.

Similar to the kernel of a linear map, the preimage of a regular value, under a smooth map $G : \mathsf{M}^m \to \mathsf{N}^n$, is a submanifold of dimension $m - n$, e.g., think of $\mathbb{S}^{n-1}$. The generalization beyond points (regular values) turns out to be remarkably useful.

Let $G : \mathsf{M}^m \to \mathsf{N}^n$ be a smooth map between smooth, boundaryless, manifolds and let $\mathsf{S}^s \subset \mathsf{N}^n$ be some smooth, boundaryless, submanifold. Then, G is said to be ***transverse*** to S^s, denoted $G \pitchfork \mathsf{S}^s$, when either $G(\mathsf{M}^m) \cap \mathsf{S}^s = \emptyset$ or

$$\mathrm{im}(DG_p) + T_{G(p)}\mathsf{S}^s = T_{G(p)}\mathsf{N}^n \tag{3.1}$$

for all $p \in G^{-1}(\mathsf{S}^s)$, see Fig. 3.1(i). Evidently but importantly, (3.1) trivially holds for G being a smooth submersion. When the transversality conditions holds, then by the *implicit function theorem*, the preimage of S^s, that is $G^{-1}(\mathsf{S}^s)$, is also a submanifold, of dimension $m - n + s$ [16, p. 28]. Two submanifolds $\mathsf{S}_1 \subseteq \mathsf{N}$ and $\mathsf{S}_2 \subseteq \mathsf{N}$ are called transverse when the inclusion map of one of them is transverse to the remaining submanifold. This boils down to the condition that $T_q\mathsf{S}_1 + T_q\mathsf{S}_2 = T_q\mathsf{N}$ for all $q \in \mathsf{S}_1 \cap \mathsf{S}_2$, which has a clear geometric interpretation. A particularly useful implication is that when $\mathsf{S}_1 \pitchfork \mathsf{S}_2$, then $\mathsf{S}_1 \cap \mathsf{S}_2$ is a submanifold itself, with $\mathrm{codim}(\mathsf{S}_1 \cap \mathsf{S}_2) = \mathrm{codim}(\mathsf{S}_1) + \mathrm{codim}(\mathsf{S}_2)$ [16, p. 30].

Generalizing transversality to maps over domains *with* a boundary, i.e., $\partial \mathsf{M} \neq \emptyset$, requires the restriction $G|_{\partial \mathsf{M}} : \partial \mathsf{M} \to \mathsf{N}$ to be also transverse to S for $G^{-1}(\mathsf{S})$ to be a manifold with boundary that satisfies $\partial\{G^{-1}(\mathsf{S})\} = G^{-1}(\mathsf{S}) \cap \partial \mathsf{M}$, see [16, p. 60], consider for example an ellipsoid in a disk as in Fig. 3.1(ii).

Then, the power of transversality is captured by the following two results.

Theorem 3.1 (Thom's (parametric) transversality theorem [16, p. 68]) *Let $H : \mathsf{T} \times \mathsf{M} \to \mathsf{N}$ be a smooth map over the manifolds T, M and N, with only M possibly having a boundary. Define the family of maps $\{G_t : t \in \mathsf{T}\}$ by $G_t(p) = H(t, p)$ and*

---

[1] This means that the corresponding critical values are of measure zero in N [13, Chap. 2]. When the condition $n \leq m$ is relaxed, one speaks of critical points when the rank of DG_p is not maximal. Sard's theorem is also a typical result aided by $C^{r>1}$ smoothness [29, Sect. 2].

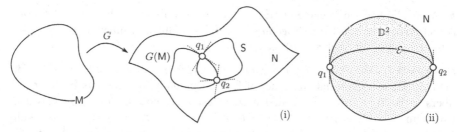

Fig. 3.1 Transversality, (i) $G : M \rightarrow N$ with $G \pitchfork S$; (ii) $\mathcal{E} \not\pitchfork \mathbb{D}^2$

let $S \subseteq N$ *be a smooth submanifold. If* $H \pitchfork S$ *and* $H|_{\partial M} \pitchfork S$, *then, for almost every* $t \in T$ *also* $G_t \pitchfork S$ *and* $G_t|_{\partial M} \pitchfork S$.

Using the language of jets, one can show that transversality generalizes regular values in the sense that transverse maps are also *generic*[2] [13, Theorem II.4.9, Corollary II.4.12]. Again, this corresponds to geometric intuition, drawing two lines at random in $\mathbb{R}^2$, they will be *almost surely* transverse. Technically, one can prove this by showing that for appropriate T, the map H is easily constructed to be a perturbation of G, yet, submersive, and hence transversal to any S.

Theorem 3.2 (Transversality homotopy extension theorem [16, pp. 72–73]) *Let G be a smooth submanifold of* N, *both without boundary, and consider a closed subset* $A \subseteq M$ *of the smooth manifold* M. *Let* $G : M \rightarrow N$ *be a smooth map such that* $G \pitchfork S$ *on* A *and* $G|_{\partial M} \pitchfork S$ *on* $A \cap \partial M$. *Then, there is a smooth map* $G' : M \rightarrow N$, *homotopic to* G *such that* $G' \pitchfork S$, $G'|_{\partial M} \pitchfork S$ *and* $G = G'$ *on a neighbourhood of* A.

By taking $A = \partial M$, Theorem 3.2 implies in particular that if $g : \partial M \rightarrow N$ is transverse to $S \subseteq N$ and g extends to M, then there is an extension $G : M \rightarrow N$ such that $G \pitchfork S$.

3.3 Bundles

Given two topological spaces E and B, the *total* and *base* space, respectively, and a continuous surjective map $\pi : E \rightarrow B$, then, the triple (π, E, B) is called a ***vector bundle*** when for each $b \in B$ the ***fiber*** $\pi^{-1}(b)$ has the structure of a real vector space, say $\mathbb{R}^k$. Moreover, for any $b \in B$, E must be *locally trivial* over some open neighbourhood U of b, that is, there is a homeomorphism $\varphi : \pi^{-1}(U) \rightarrow U \times \mathbb{R}^k$. Additionally, φ should preserve the base and fiber structure, i.e., for $\pi_U(U, \mathbb{R}^k) = U$, $\pi_U \circ \varphi = \pi$ and for each $b' \in U$, $\varphi(\pi^{-1}(b'))$ is linearly isomorphic to $\mathbb{R}^k$. Given a vector bundle $\pi : E \rightarrow B$, a *section* is a continuous map $\sigma : B \rightarrow E$ such that

---

[2] Meaning, with respect to the C^∞ (Whitney) topology.

$\pi \circ \sigma = \text{id}_B$. Sections, denoted $\Gamma(E)$, will aid in describing feedback laws later on. A section of interest is the **zero section** $Z_\pi \in \Gamma(E)$, defined by mapping $b \in B$ to the zero element of the fiber $\pi^{-1}(b)$, i.e., $Z_\pi(B) \simeq_t B$, see also Fig. 3.2(i).

Example 3.1 (*Vector bundle retraction*) Consider a vector bundle $\pi : E \to B$, then $Z_\pi(B)$ is a deformation retract of E. Conceptually, one retracts E along the fibers to B. To show this, we first need to establish how to *transition* between two homeomorphisms $\varphi : \pi^{-1}(U) \to U \times \mathbb{R}^k$ and $\psi : \pi^{-1}(V) \to V \times \mathbb{R}^k$ with $U \cap V \neq \emptyset$. It follows from the structure preservation that $\pi_{U \cap V} \circ \varphi \circ \psi^{-1} = \pi_{U \cap V}$ and as such for any $b \in U \cap V$ we have $\varphi \circ \psi^{-1}(b, x) = (b, g(b, x))$ for some $g : (U \cap V) \times \mathbb{R}^k \to \mathbb{R}^k$. By the properties of φ, ψ, the map $x \mapsto g(b, x)$ must be a linear bijection, that is, $g(b, x) = A(b)x$ for $A(b) \in \text{GL}(k, \mathbb{R})$. Evidently, this means that the transition $\varphi \circ \psi^{-1}$ is linear in $x \in \mathbb{R}^k$. Now, as we can let E be locally trivial over some neighbourhood U of $b \in B$, construct, in local coordinates the homotopy $H(t, (b, x)) = (t, (b, (1 - t)x))$. As we just saw, the local transition maps are also linear in x, as such, this construction is well-defined over the entire vector bundle and indeed yields a deformation retract from E onto $Z_\pi(B)$.

Example 3.1 also shows why vector bundles admit zero sections; the transition $\varphi \circ \psi^{-1}$ maps 0 to 0. For more information, see [18, Chap. 4].

Then, to characterize neighbourhoods of embedded submanifolds, it is useful to introduce the following. The vector bundle $\pi_S : S \to B$ is a **subbundle** of the vector bundle $\pi : E \to B$ when $S \subseteq E$, $\pi_S = \pi|_S$ and for all $b \in B$ one has that $\pi_S^{-1}(b) = S \cap \pi^{-1}(b)$ is a linear subspace of the fiber $\pi^{-1}(b)$. A subbundle of particular interest is the **normal bundle** of an embedded (or immersed) submanifold $M^n \subseteq \mathbb{R}^d$, denoted $TM^\perp$ or NM. This bundle is the orthogonal complement, under the Euclidean inner-product inherited from $\mathbb{R}^d$, of the tangent bundle in the embedding space. In particular, let $S \subseteq M$ be an embedded submanifold of a smooth manifold M. For simplicity assume M is itself embedded into Euclidean space. Then, the normal bundle $TS^\perp$, with respect to TS, is given by $TS^\perp = \sqcup_{s \in S} T_s S^\perp$, for $T_s S^\perp$ the orthogonal complement, with respect to the Euclidean metric, to $T_s S$. Algebraically, $TS^\perp \subseteq TM$ is given by the quotient $TM|_S / TS$ [18, Chap. 4.2].

Now, following [27], consider some embedded submanifold $M \subseteq \mathbb{R}^d$ and define the map $w : NM \to \mathbb{R}^d$ by $w(p, n) = p + n$. Additionally, define the set $V = \{(p, n) \in NM : \|n\| < \delta(p)\}$ for some continuous function $\delta : M \to \mathbb{R}_{>0}$ such that V is open. Then, a neighbourhood U of M in $\mathbb{R}^d$ that is diffeomorphic to $w(V)$ is said to be a **tubular neighbourhood** of M. The *Tubular neighbourhood theorem* states that every embedded submanifold of $\mathbb{R}^d$ has a tubular neighbourhood [27, Theorem 6.24]. By exploiting the diffeomorphism $w : V \to U$, one can show the following, see also the ϵ-neighbourhood theorem [16, p. 69] and Fig. 3.2(ii).

Proposition 3.1 (Tubular neighbourhood retraction [27, Proposition 6.25]) *If U is a tubular neighbourhood of some smooth embedded submanifold $M \subseteq \mathbb{R}^d$, there is a smooth map $r : U \to M$ that is both a retraction and a submersion.*

Proposition 3.1 will be useful in a later stage, it also allows for showing the following well-known result by Whitney, here, G is understood to be δ-close to F.

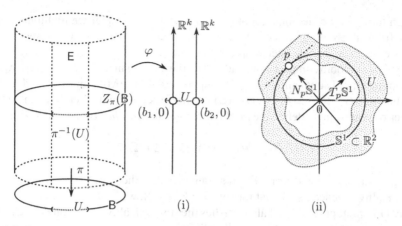

Fig. 3.2 Bundles, (i) the vector bundle $\pi : E \rightarrow B$ with zero section $Z_\pi(B)$; (ii) a tubular neighbourhood U of $\mathbb{S}^1 \subset \mathbb{R}^2$

Theorem 3.3 (Whitney's Approximation Theorem [27, Theorem 6.26]) *Let* M *and* N *be smooth manifolds, with only* M *possibly having boundary. For any continuous map* $F : M \rightarrow N$ *there is a smooth map* $G : M \rightarrow N$ *homotopic to* F.

Indeed, G can even be chosen such that $G \pitchfork S$ for any $S \subset N$ [27, Theorem 6.36].

In what follows we mostly study continuous maps on smooth manifolds. However, most of the results at our disposal require some degree of smoothness to be proven, not merely continuity. At the same time, most of these results are invariant under homotopy. As such, Theorem 3.3 allows from bridging the gap between continuity and smoothness, which would be otherwise non-trivial. See also [18, Lemma 5.1.5].

We end this section with a comment on a generalization of vector bundles. Instead of demanding that the fibers are vector spaces, one can relax this to the demand that $\pi^{-1}(b)$ is homeomorphic to some topological space F, while E must still be locally trivial. In this case, the 4-tuple (π, E, B, F) represents what is called a *fiber bundle*, e.g., $(\pi, \mathbb{S}^3, \mathbb{S}^2, \mathbb{S}^1)$ is arguably the most influential example and is called the *Hopf fibration*. Importantly, for fiber bundles the existence of a continuous section is not immediate. A section does exist when F is contractible [39, Part III]. We return to this topic in Chap. 6. Additionally, one can specify the *structure group*, e.g., instead of $GL(k, \mathbb{R})$ in Example 3.1, one could consider $O(k, \mathbb{R})$ and so forth.

3.4 Intersection and Index Theory

The practical classification of manifolds and maps over these manifolds relies on *topological invariants* (frequently, *homotopy invariants*). Using the previous results on transversality we provide a brief overview of the construction of the key topological invariant for this work: *the Euler characteristic*. In this chapter this in done through the lens of differential topology and in the next chapter we highlight arguments from algebraic topology. We point out that the material in this section is instrumental to appreciate later chapters and sections.

Let M be a smooth compact manifold and let the smooth map $G : M \to N$ be transverse to the closed submanifold $S \subseteq N$. Suppose that $\dim(M) + \dim(S) = \dim(N)$ such that $\dim(G^{-1}(S)) = 0$, i.e., $G^{-1}(S)$ is a finite set of points. Let $\#(\cdot)$ denote the number or points, then, define the ***mod 2 intersection number*** of the pair (G, S) as

$$I_2(G, S) = \#(G^{-1}(S)) \bmod 2 \in \mathbb{Z}/2\mathbb{Z}. \tag{3.2}$$

For a general map, recall from Thom's transversality theorem (Theorem 3.1) that transversality is generic such that for any $G' : M \to N$, we let $I_2(G', S) = I_2(G, S)$ for any G homotopic to G' that also satisfies the transversality condition $G \pitchfork S$. The following result shows why this is well-defined.

Theorem 3.4 (Mod 2 intersection homotopy invariance [16, p. 78]) *Let M^m be compact and let $S^s \subseteq N^n$ be a closed submanifold such that $m + s = n$, all manifolds being boundaryless and smooth. Then, for any pair of smooth maps $G_1, G_2 : M^m \to N^n$ being homotopic, one has $I_2(G_1, S^s) = I_2(G_2, S^s)$.*

As this result exemplifies upcoming material, a proof from [16] is collected.

Proof By definition we have $I_2(G', S^s) = I_2(G, S^s)$ such that $G \simeq_h G'$ and $G \pitchfork S^s$. Hence, by the transitive property of homotopies, without loss of generality, let $G_1 \pitchfork S^s$ and $G_2 \pitchfork S^s$. Then, let $H : [0, 1] \times M^m \to N^n$ be the homotopy between G_1 and G_2. By construction, $H|_{\partial\{[0,1]\times M^m\}}$ is transverse to S^s. By the homotopy transversality extension theorem (Theorem 3.2) we can assume that $H \pitchfork S^s$, i.e., as the map can be extended. This implies that $H^{-1}(S^s)$ is a one-dimensional manifold with boundary, defined via $\partial\{H^{-1}(S^s)\} = (\{0\} \times G_1^{-1}(S^s)) \cup (\{1\} \times G_2^{-1}(S^s))$. Then the result follows by observing that one-dimensional manifolds have boundaries with an even number of points, motivating the definition of $I_2(\cdot, \cdot)$, see [29, Appendix]. $\qquad\blacksquare$

For example, if we construct the constant map $C : M^m \to N^n$, for $n > 0$, defined by $C : p \mapsto q'$ for all $p \in M^m$ and some $q' \in N^n \setminus S^s$, then the transversality condition (3.1) holds trivially and indeed $I_2(C, S^s) = 0$. We point out that exactly results like Theorem 3.4 are the reason why we discussed (transversality in the context of) manifolds *with* boundaries, i.e., see that the manifolds under consideration in that theorem are all boundary*less* themselves while in the proof we exploit $[0, 1] \times M^m$.

Again, one can use the inclusion map to define the intersection number between manifolds. Let both $S_1 \subseteq N$ and $S_2 \subseteq N$ be compact and boundaryless, and of ***complementary dimension***, that is $\dim(S_1) + \dim(S_2) = \dim(N)$. Then, $I_2(S_1, S_2) = I_2(\iota_{S_1}, S_2)$, for $\iota_{S_1} : S_1 \hookrightarrow N$. If $S_1 \pitchfork S_2$, then, by construction $I_2(S_1, S_2) = \#(\iota_{S_1}^{-1}(S_2)) \bmod 2$, which is simply the number of intersection points, modulo 2. Due to the homotopy invariance, when $I_2(S_1, S_2) = 1$, any manifold homotopic to S_1 intersects S_2, e.g., consider two circles on the torus. In that sense, $I_2(\cdot, \cdot)$ is robust.

Lemma 3.1 (Compact manifolds are generally not contractible) *Compact, bound-aryless manifolds N^n with $n \geq 1$ are not contractible.*

Proof Suppose N^n is contractible, then id_{N^n} is homotopic to some constant map $C : q \mapsto q'$ over N^n for some $q' \in N^n$. Then, given any compact M^m and closed sub-manifold $S^s \subseteq N^n$ such that $m + s = n$, pick any smooth map $G : M^m \to N^n$ transverse to S^s. It follows by homotopy equivalence and composition that $I_2(G, S^s) = I_2(\mathrm{id}_{N^n} \circ G, S^s) = I_2(C \circ G, S^s) = 0$ (if needed, perturb S^s not to contain q'). In particular, consider the setting of G being the identity map on N^n, forcing S^s to be a point, i.e., $s \in N^n \setminus \{q'\}$. One is led to a contradiction as $N^n \cap S^s = \{s\}$, yet, $I_2(\mathrm{id}_{N^n}, S^s) = 0$. For a reference, see [16, Exercises 5–6, p. 83].

Manifolds of dimension 0 can be contractible and indeed, the proof fails for $n = 0$ as in that case $I_2(C \circ G, S^s)$ is not necessarily 0. When $\partial N^n \neq \emptyset$ contractability might hold. In that case, the proof fails as the homotopy invariance of I_2 does not necessarily carries over, e.g., consider the unit interval moving through to the unit circle.[3] One can also employ *Poincaré duality* (see Sect. 4) to show Lemma 3.1.

We introduce one more concept. Let $G : M^m \to N^n$ be a smooth map from a compact to a connected manifold with $m = n$, both boundaryless.[4] Then, for any $q \in N^n$, the number $I_2(G, \{q\})$ is the **mod 2 degree** of G, denoted $\deg_2(G)$. Note that this number is a homotopy invariant by Theorem 3.4 and it is the same for any $q \in N^n$ [16, p. 81]. This means that given a regular value q, $\deg_2(G) = \#(G^{-1}(q))$ mod 2.

An obstruction related to Lemma 3.1 holds for deformation retracts.

Lemma 3.2 (Compact manifolds admit no proper deformation retract) *Let M be a boundaryless, compact, connected, manifold. Then, there is no proper subset A of M such that M deformation retracts onto A.*

Proof For the sake of contradiction, assume there would be such a deformation retract, let $r : M \to A$ be the retraction and let $\iota_A : A \hookrightarrow M$ be the inclusion map. Now clearly, $\deg_2(\iota_A \circ r) = 0$ as one can consider the preimage of any point in $M \setminus A$. However, by assumption $\iota_A \circ r \simeq_h \mathrm{id}_M$, such that $0 = \deg_2(\iota_A \circ r) = \deg_2(\mathrm{id}_M) \neq 0$.

For example, by Lemma 3.2, the group $SO(3, \mathbb{R})$ cannot deformation retract onto $SO(2, \mathbb{R}) \hookrightarrow SO(3, \mathbb{R})$. Spot again the boundaryless assumption.

Due to the binary evaluation, however, the insights gained from mod 2 intersection theory are limited. Endowing a space with an *orientation* allows for a different manipulation of $\#G^{-1}(S)$ with far reaching ramifications.

A smooth manifold M^m is said to be **oriented** when an admissible smooth orientation is selected (see below). All orientations will be with respect to the *standard*

---

[3] We stay in the realm of topological manifolds, often assumed to be compact and without boundary. Relaxing these assumptions often requires the homotopy to be *proper*. A map $G : X \to Y$ is said to be **proper** when $G^{-1}(K)$ is compact for any compact $K \subseteq Y$. The intuition is that under this properness assumption, $G^{-1}(q)$ is a compact set indeed for any point $q \in Y$, which is exploited in a variety of manners, e.g., when working with the *degree*. For more, see [33, Chap. III].

[4] One can handle non-trivial boundaries for instance when G is proper and such that $G : \partial M^m \to \partial N^n$, or using axiomatic degree theory [33, Chap. IV], which we only briefly mention.

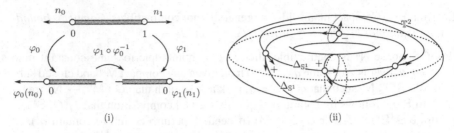

Fig. 3.3 Oriented intersections, (i) orientation on a one-dimensional manifold with boundary; (ii) self-intersection number $\chi(\mathbb{S}^1) = I(\Delta_{\mathbb{S}^1}, \Delta_{\mathbb{S}^1}) = 0$

orientation on $\mathbb{R}^m$. Given a vector space V^m, if a basis B for V^m is isomorphic to $\mathbb{R}^m$ by means of an *orientation-preserving* map, that is, a linear map with strictly positive determinant, then V^m is said to be *positively oriented* under B. Otherwise, V^m is *negatively oriented*. For manifolds with boundary, the orientation on the boundary is the one induced by the outward normal. For 1-dimensional manifolds with boundary, the domain of the local coordinates needs to be altered for this to work, that is, allow for mapping to $\mathbb{R}_{\leq 0}$. Following [41, Example 21.8], given two charts (U_0, φ_0) and (U_1, φ_1) on $[0, 1]$ defined by $U_0 = [0, 1)$, $\varphi_0(x) = x$ and $U_1 = (0, 1]$, $\varphi_1(x) = x - 1$, observe that $\varphi_1 \circ \varphi_0^{-1} = x - 1$ and $\varphi_0 \circ \varphi_1^{-1} = x + 1$, as such, $[0, 1]$ is *orientable*, that is, the transition functions preserve orientation. This does, however, mean that the outward induced orientation assigns -1 to the point 0 and $+1$ to the point 1 as shown in Fig. 3.3(i). This example will also aid in illustrating why the (oriented) degree is defined for boundary*less* manifolds below.

It turns out that transversality naturally leads to an orientation on the manifold of interest. Let $G : \mathsf{M} \to \mathsf{N}$ be smooth, with $\mathsf{S_N} \subseteq \mathsf{N}$, N and $\mathsf{S_N}$ boundaryless, $G \pitchfork \mathsf{S_N}$, $G|_{\partial \mathsf{M}} \pitchfork \mathsf{S_N}$, and $\mathsf{M}, \mathsf{N}, \mathsf{S_N}$ all oriented. Let $\mathsf{S_M} = G^{-1}(\mathsf{S_N})$ and define $T_p^{\perp}(\mathsf{S_M}; \mathsf{M})$ to be complementary to $T_p \mathsf{S_M}$, that is, (3.3a) below must be satisfied, e.g., $T_p^{\perp}(\mathsf{S_M}; \mathsf{M})$ is the orthogonal complement when a metric is defined (which is irrelevant for the proceeding discussion). By combining transversality (3.1) with $T_p \mathsf{S_M}$ being the preimage of $T_{G(p)} \mathsf{S_N}$ *under* DG_p, we obtain for all $p \in G^{-1}(\mathsf{S_N})$ that

$$T_p^{\perp}(\mathsf{S_M}; \mathsf{M}) \oplus T_p \mathsf{S_M} = T_p \mathsf{M} \tag{3.3a}$$

$$DG_p T_p^{\perp}(\mathsf{S_M}; \mathsf{M}) \oplus T_{G(p)} \mathsf{S_N} = T_{G(p)} \mathsf{N}. \tag{3.3b}$$

Then, as by transversality, the kernel of DG_p must be contained in $T_p \mathsf{S_M}$ (in the preimage of $T_{G(p)} \mathsf{S_N}$), (3.3b) induces an orientation on $T_p^{\perp}(\mathsf{S_M}; \mathsf{M})$ and subsequently, via (3.3a) an orientation on $T_p \mathsf{S_M}$, called the **preimage-orientation**.

Now the **oriented intersection number** is defined similar as I_2, yet we add up orientation numbers, with respect to the preimage-orientation, of all $p \in G^{-1}(\mathsf{S}) = \mathsf{S_M}$, denoted $I(G, \mathsf{S})$. In particular, let $G : \mathsf{M} \to \mathsf{N}$ be smooth and transverse to $\mathsf{S} \subseteq \mathsf{N}$. Under compactness of M, closedness of S and a complementary dimension condition, $G^{-1}(\mathsf{S})$ is a finite set of points. Now also assume that M, N and S are oriented. As $G \pitchfork S$ one has for any $p \in G^{-1}(S)$ that $DG_p T_p M \oplus T_{G(p)}S = T_{G(p)}N$. It follows from (3.3) that the orientation number at p is defined and equals $+1$ when the orientation on both sides of the equation agrees, whereas the orientation number is -1 when they do not. Note that 0-dimensional manifolds also have an orientation attached to them and that (3.3a) only implies that $T_p^{\perp}(p; M)$ and $T_p M$ are isomorphic.

Most importantly, oriented intersection numbers are also homotopy invariant [16, p. 108]. Indeed, this result is similar to Theorem 3.4 and to that end we clarified above how to define orientation on a boundary. More specifically, in the mod 2 intersection setting, the proof of Theorem 3.4 relied on the sum of boundary points of one-dimensional manifolds, modulo 2, always being 0. Recalling Fig. 3.3, in the oriented case, the sum of the orientation numbers of exactly those points is also always 0. Given this homotopy invariance, we define the oriented intersection number of general smooth maps G' via $I(G', S) = I(G, S)$ for a map G that does satisfy the aforementioned conditions and is homotopic to G'. Exploiting this observation regarding one-dimensional manifolds one can show the following.

Lemma 3.3 (Intersection number extension lemma [16, p. 108]) *Let $g : \partial M \to N$ be a smooth map transverse to a closed, smooth, boundaryless submanifold $S \subseteq N$ of complementary dimension. If g extends to the entire smooth, compact manifold M, then $I(g, S) = 0$.*

Proof (*sketch*) Let $G : M \to N$ be the extension, by construction $G|_{\partial M} \pitchfork S$ and by Theorem 3.2 one can take $G \pitchfork S$ such that $G^{-1}(S) \subseteq M$ is a one-dimensional manifold. The result follows by recalling that by transversality $\partial \{G^{-1}(S)\} = g^{-1}(S)$.

Similar as before, for any smooth map between boundaryless manifolds $G : M^m \to N^n$ with $m = n$, M^m compact and N^n being connected,[5] we define the **degree** of G as $\deg(G) = I(G, \{q\})$ for any $q \in N^n$ (recall Theorem 3.1). In this case the orientation number can be computed using the same reasoning as before, i.e., for $p \in G^{-1}(q)$ we check if DG_p will *preserve* or *reverse* the orientation on $T_p M^m$. If M^m is endowed with the canonical positive orientation, then, for any regular value q

$$\deg(G) = \sum_{p \in G^{-1}(q)} \operatorname{sgn} \det(DG_p). \tag{3.4}$$

The form of (3.4) goes back to Kronecker and the intuition of (3.4) is that the number $\deg(G)$ represents how many times (net) the domain "*wraps around*" the codomain under the map G. See also Sect. 4 for the homological viewpoint.

Example 3.2 (*Homotopies and the n-sphere*) Consider the n-sphere $\mathbb{S}^n \subset \mathbb{R}^{n+1}$, a smooth manifold X and a smooth map $G : X \to \mathbb{S}^n$. Now, let $G' : X \to \mathbb{S}^n$ be such that $\|G(p) - G'(p)\|_2 < 2$ for all $p \in X$. This condition implies that G is homotopic to G' as $H : [0, 1] \times X \to \mathbb{S}^n$ defined by

$$H(t, p) = \frac{(1 - t)G(p) - tG'(p)}{\|(1 - t)G(p) - tG'(p)\|_2}$$

---

[5] Without the *connectedness* assumption the degree might differ from component to component.

is the corresponding homotopy. This construction concurrently shows the *robustness* of $\deg(\cdot)$. Note, the $\| \cdot \|_2$ condition is not necessary as on $\mathbb{S}^1 \subset \mathbb{R}^2$ one can rotate the identity map to its negation. In fact, *Hopf's degree theorem* states that if M is compact, connected, boundaryless and oriented, then any continuous map $g : M \to \mathbb{S}^n$ is homotopic to $g' : M \to \mathbb{S}^n$ if and only if $\deg(g) = \deg(g')$ [29, Sect. 7].

As before, given two submanifolds S_1 and S_2, we can also define $I(S_1, S_2)$ via their respective inclusion maps. Note, however, that by no means $I(S_1, S_2)$ is necessarily equal to $I(S_2, S_1)$. In particular, one can show [16, pp. 113–115] that if S^s and R^r are compact submanifolds of N^n of complementary dimension, then,

$$I(S^s, R^r) = (-1)^{s \cdot r} I(R^r, S^s). \tag{3.5}$$

We can now define the central invariant of this work. Let Δ_V denote the *diagonal* of $V \times V$, that is, $\Delta_V = \{(v, v) : v \in V\}$, and let M be a smooth, boundaryless, compact and orientable manifold, then, its **Euler characteristic** is defined as

$$\chi(M) = I(\Delta_M, \Delta_M). \tag{3.6}$$

Note, (3.5) implies that $\chi(M^m) = 0$ when m is odd. One should interpret the self-intersection number (3.6) with the aforementioned transversality conditions and homotopy invariance taken into account. For example, for $M = \mathbb{S}^1$, think of $\Delta_{\mathbb{S}^1}$ as a particular circle on $\mathbb{S}^1 \times \mathbb{S}^1 = \mathbb{T}^2$. Then, $\chi(\mathbb{S}^1)$ captures to what extent a homotopy of $\Delta_{\mathbb{S}^1}$ remains entangled with $\Delta_{\mathbb{S}^1}$, see Fig. 3.3(ii). See also Sect. 4.2 for $\chi(M)$ through the lens of algebraic topology. We will follow a constructive approach in showing how $\chi(M)$ relates to qualitative properties of vector fields on M. We do not start with vector fields, but with maps. This simplifies the analysis and in contrast to combinatorial/algebraic approaches, this makes it possible to relate some upcoming material to discrete-time systems via time-one maps cf. Section 8.1.

Let $G : M \to M$ be a smooth map over a smooth, boundaryless, compact, orientable manifold M. The existence of fixed points of G is for instance captured by the **Lefschetz number** $L(G) = I(\Delta_M, \mathrm{graph}(G))$ being different from 0. Indeed, $L(\mathrm{id}_M) = \chi(M)$. A map G is called a **Lefschetz map** when $\mathrm{graph}(G) \pitchfork \Delta_M$, yielding robust fixed-point properties. Now, one can derive that G being Lefschetz over M^m is equivalent to $DG_p - I_m$ being invertible (the reader is invited to visualize this). Fixed points $p \in M^m$ of G such that $DG_p - I_m$ is invertible are called **Lefschetz fixed points** and for Lefschetz maps one can compute $L(G)$ via **local Lefschetz numbers**, that is $L(G) = \sum_{p=G(p)} L_p(G)$, where the sign of the corresponding isomorphism defines the local Lefschetz numbers, i.e., the orientation numbers of the Lefschetz fixed point. By comparing orientations, one can show the following.

Proposition 3.2 (Lefschetz number of Lefschetz fixed point [16, p. 121]) *Let $G : M^m \to M^m$ be a smooth Lefschetz map over a smooth, boundaryless compact, orientable manifold M^n, then*

$$L(G) = \sum_{p=G(p)} \operatorname{sgn} \det(DG_p - I_m). \tag{3.7}$$

Equation (3.7) is appealing, but only valid for Lefschetz fixed points. In the dynamical systems context, by only considering Lefschetz fixed points we are ignoring a set of *structurally unstable* fixed points. To make sure that $L(G)$ is well-defined, and computable, we need to refine the notion of $L_p(G)$ for generic maps.

We start in the Euclidean setting. Let $G : \mathbb{R}^m \to \mathbb{R}^m$ be smooth with a fixed point $p \in \mathbb{R}^m$ and define for some closed neighbourhood cl $U \subseteq \mathbb{R}^m$ around p, containing no other fixed points, the map $g : \partial U \to \mathbb{S}^{m-1}$ by

$$g : q \mapsto g(q) = \frac{G(q) - q}{\|G(q) - q\|_2}. \tag{3.8}$$

Then, let the *generalized local Lefschetz number* $\widetilde{L}_p(G)$ be equal to the degree of this map, that is $\widetilde{L}_p(G) = \deg(g)$. For this construction to be of any use, the degree must be invariant under a change of neighbourhood U. Pick any other closed neighbourhood cl U', strictly contained in cl U, then, as g extends to cl $U \setminus U'$, the degree vanishes on the boundary of this set by means of Lemma 3.3. However, by construction, this implies that the degree under both neighbourhoods must be equal (as the induced orientations must be the opposite), see also [16, p. 127]. If cl U and cl U' merely have a non empty intersection, then, one first finds a larger closed neighbourhood cl U'' that contains both sets and the previous argument extends. Moreover, it can be shown that when p is a Lefschetz fixed point $\widetilde{L}_p(G) = L_p(G)$ [16, p. 128]. The following result is instrumental in linking local Lefschetz numbers.

Proposition 3.3 (On local Lefschetz numbers [16, pp. 126–129]) *Let the smooth map $G : \mathbb{R}^m \to \mathbb{R}^m$ have some isolated fixed point $p^\star$ and let $\mathbb{B}^m$ be an open ball containing $p^\star$ such that cl $\mathbb{B}^m$ does not contain any other fixed points of G. Next, pick a map G' that equals G outside of some compact subset of $\mathbb{B}^m$ and has all its fixed points in $\mathbb{B}^m$ being of the Lefschetz type. Then, the pair (G, G') satisfies $\widetilde{L}_{p^\star}(G) = \sum_{p=G'(p)} L_p(G')$ for $p \in \mathbb{B}^m$.*

Proof *(sketch)* First, $\widetilde{L}_{p^\star}(G)$ equals the degree of the map $g : \partial \mathbb{B}^m \to \mathbb{S}^{m-1}$ defined via (3.8). By construction, on $\mathbb{R}^m \setminus \mathbb{B}^m$, G can be replaced with G'. Let $p_1, \ldots, p_k$ be the set of Lefschetz fixed points of G' in $\mathbb{B}^m$ and let $(\mathbb{B}^m_{r_i})_i \subseteq \mathbb{B}^m$ be a disjoint set of sufficiently small balls around those points such that $\partial \mathbb{B}^m \cap (\cup_i \mathbb{B}^m_{r_i}) = \emptyset$. Let $\mathbb{B}' = \mathbb{B}^m \setminus \cup_i \mathbb{B}^m_{r_i}$, considering (3.8) for G' over $\partial \mathbb{B}' \to \mathbb{S}^{n-1}$, by construction this map extends to $\mathbb{B}'$ such that by Lemma 3.3 the degree of this map must equal 0. Then the claims follows by observing that $\partial \mathbb{B}'$ consists out of $\partial \mathbb{B}^m$ and $\partial \{\cup_i \mathbb{B}^m_{r_i}\}$, with for the latter set(s) the orientation being flipped with respect to $\mathbb{B}^m$.

Now, given a general smooth map $G : M^m \to M^m$ with isolated fixed point $p^\star$, let $\psi : U \to \mathbb{R}^m$ be a diffeomorphism around $p^\star$ mapping $p^\star$ to 0 and consider $\psi \circ G \circ \psi^{-1} : \mathbb{R}^m \to \mathbb{R}^m$. First, assume that $p^\star$ is a Lefschetz fixed point, hence, $DG_{p^\star} - I_m$ is an isomorphism. See that in coordinates one has $D(\psi \circ G \circ \psi^{-1})_0 - I_m =$

$D\psi_{p^*} \circ (DG_{p^*} - I_m) \circ D\psi_0^{-1}$ such that the local Lefschetz number is preserved. For generic fixed points, employ Proposition 3.3. Homotopy invariance, generality of Lefschetz maps and Propositions 3.2–3.3 lead to the following generalization.

Theorem 3.5 (General Lefschetz numbers [16, p. 130], [17, Sect. 2.C]) *Let* G : $M^m \to M^m$ *be a smooth map over a smooth, boundaryless, compact, oriented manifold* M^m *with finitely many fixed points. Then,*

$$L(G) = \sum_{p=G(p)} \tilde{L}_p(G). \tag{3.9}$$

Remark 1 (Axiomatic degree theory [9, Appendix B], [33, Chap. IV]) Degree theory on closed manifolds is powerful, yet sometimes restrictive. It turns out that the concept can be generalized axiomatically to closed subsets of $\mathbb{R}^n$. Inspired by the properties of $\deg(\cdot)$ one can *derive* a map $d(G, D, q)$ with these desirable features like (3.4), where now $G : D \to \mathbb{R}^n$ is a $C^{r \geq 0}$ map over some bounded set $D \subset \mathbb{R}^n$ and q is a regular value such that $G^{-1}(q) \notin \partial D$. We will not appeal to this construction, but regarding further reading this is important to be aware of.

3.5 Poincaré–Hopf and the Bobylev–Krasnosel'skiĭ theorem

Lefschetz fixed point theory allows for analyzing flows and discrete-time dynamical systems, however, we are ultimately interested in continuous-time dynamical systems and particularly vector fields. The reason being, first principles possibly provide one with a differential equation approximating[6] some phenomenon, having access to an explicit solution (flow), however, is rare. Hence, we switch from maps to vector fields on M, where the set of C^r-smooth vector fields on M is denoted by $\mathfrak{X}^r(M)$, see Sect. 5.1 for more details on continuous-time dynamical systems.

We will start again on $\mathbb{R}^n$ and define the vector field analogue of the local Lefschetz number, as introduced by Kronecker/Poincaré and formalized by Hopf.

Definition 1 (Index of a zero) Consider some open set $\Omega \subseteq \mathbb{R}^n$ and let $p^* \in \Omega$ be an isolated zero of the smooth vector field $X \in \mathfrak{X}^\infty(\Omega)$. Let U be a neighbourhood of p^* such that p^* is the only zero of X over cl U and define the map $v : \partial U \to \mathbb{S}^{n-1}$ by $v : p \mapsto X(p)/\|X(p)\|_2$. Then, the index of p^* is defined as

$$\mathrm{ind}_{p^*}(X) = \deg(v). \tag{3.10}$$

Indeed, if one is not aware of $X \in \mathfrak{X}^\infty(\Omega)$ having a zero on $U \subseteq \Omega$, index computations provide a partial answer. To lift the construction to manifolds, one can show

---

[6] We like to already emphasize that most tools discussed in this work are particularly suitable when models are only *roughly* known and one is after insights into qualitative behaviour.

that (3.10) is invariant under diffeomorphisms [29, p. 33]. This will be shown after establishing the link between vector field indices and local Lefschetz numbers.

Proposition 3.4 (Vector field indices and Lefschetz numbers [16, pp. 135–136]) *Let X be a smooth vector field over some open neighbourhood $\Omega \subseteq \mathbb{R}^n$ of the origin, only vanishing at 0. Let $\{\varphi_X^t : t \geq 0\}$ be a family of mutually homotopic maps, smoothly mapping Ω to itself, with $\varphi_X^0 = I_n$ and for $t \neq 0$, φ_X^t having no fixed points besides 0. If $X(p)$ is tangent to $\varphi_X^t(p)$ at $t = 0$ for all $p \in \Omega$, then*

$$\mathrm{ind}_0(X) = \tilde{L}_0(\varphi_X^t). \tag{3.11}$$

Proof (*sketch*) First, by direct integration, one can show that if $g(t)$ is a smooth function, then there is a smooth function $r(t)$ such that $g(t) = g(0) + t(d/ds)g(s)|_{s=0} + t^2 r(t)$. Then, fix $p \in \Omega$ and apply this coordinate-wise to $\varphi_X^t(p)$ as seen as function in t, that is $\varphi_X^t(p) = \varphi_X^0(p) + t(d/ds)\varphi_X^s(p)|_{s=0} + t^2 R(t, p)$, for $R(t, p)$ the vector-valued remainder term. Rearranging yields $\varphi_X^t(p) - p = tX(p) + t^2 R(t, p)$. By construction, for $p \neq 0$ and $t \neq 0$ we have $\varphi_X^t(p) - p \neq 0$ and as such

$$\frac{\varphi_X^t(p) - p}{\|\varphi_X^t(p) - p\|_2} = \frac{X(p) + tR(t, p)}{\|X(p) + tR(t, p)\|_2} \tag{3.12}$$

is well-defined. The left part of (3.12) can be identified with $\tilde{L}_0(\varphi_X^t)$, whereas the right part defines a homotopy in t. For $t = 0$ we recover $\mathrm{ind}_0(X)$ and as the degree is homotopy invariant the result follows (only use this last argument on the right). $\square$

Proposition 3.4 is of interest in its own right, but in particular, to show the following invariance. Let X be a smooth vector field over M with some isolated equilibrium point $p^\star$. Let U be a neighbourhood of $p^\star$ and let ψ be a diffeomorphism from U onto a neighbourhood V of 0. As such, the *pushforward* $D\psi \circ X \circ \psi^{-1} = \psi_* X$ defines a vector field in local coordinates. In particular, we have

$$\mathrm{ind}_0(\psi_* X) = \tilde{L}_0(\varphi_{\psi_* X}^t) = \tilde{L}_{p^\star}(\varphi_X^t) = \mathrm{ind}_{p^\star}(X), \tag{3.13}$$

where we exploit that M can be embedded in some Euclidean space. Most importantly, this shows that the index is well-defined on manifolds. Also see that (3.10) is purely local and does not rely on M being orientable. See Section 5.1 for the formal introduction of the pushforward of a vector field X under a smooth map ψ.

Given the aforementioned discussion, let a vector field X, with only isolated zeroes, give rise to a flow φ_X^t. Indeed, the fixed points of this flow are the zeros of X. Moreover, for sufficiently small t, φ_X^t behaves similar to the identity map. In this case the Lefschetz number collapses to the Euler characteristic $\chi(M)$; this is the Poincaré–Hopf theorem, named after its initiator and key contributor.

Theorem 3.6 (Poincaré–Hopf theorem [29, p. 35]) *Let M be a smooth, compact, oriented, boundaryless manifold. Then, for any smooth vector field $X \in \mathfrak{X}^\infty(M)$ with only isolated zeroes $\{p_i^\star\}_{i \in \mathcal{I}} \subset M$ one has*

$$\chi(M) = \sum_{i \in \mathcal{I}} \mathrm{ind}_{p_i^*}(X). \tag{3.14}$$

Proof Hopf preferred a combinatorial/algebraic approach [19, Chap. 1], we, however, follow [16, p. 137], embed M in some Euclidean space $\mathbb{R}^d$ (which we can do with $d \leq 2\mathrm{dim}(M) + 1$ by our topological assumptions on M [27, Chap. 6]) and construct a tubular neighbourhood U of M. Then, by Proposition 3.1 we know we can find a U such that the normal projection $\pi : U \to$ M is a C^∞ retraction. As M is compact, $p + tX(p)$ will be contained in U for sufficiently small $t > 0$ and any $p \in$ M. Then, construct the map $\varphi^t(p) = \pi(p + tX(p))$. For any $p \in$ M we have

$$\frac{\mathrm{d}}{\mathrm{d}t}\varphi^t(p)|_{t=0} = X(p),$$

such that Theorem 3.5 and Proposition 3.4 apply if we can show that the fixed points of φ^t equal the equilibrium points of X. Any zero of X leads to a fixed point of φ^t. Since π is a normal projection, any other fixed point must have $tX(p)$ being perpendicular to T_pM, which implies that $X(p)$ must be 0. Then as φ^0 is id_M and homotopic to φ^t (for sufficiently small $t > 0$), the result follows as $L(\mathrm{id}_M) = I(\Delta_M, \Delta_M) = \chi(M)$.

It is known that the Poincaré–Hopf theorem can be used to assess if vector fields have unique equilibria over compact sets, early remarks of this nature can be found in [36, 43]. In the context of dynamical control systems, similar tools have been used in [2, 3, 7, 8, 12, 21, 32, 37, 42] and more recently in [20, 24, 25, 44]. This approach can also be seen in the context of economics [10, 28], quantum mechanics [1] and optimization [38]. We will also exploit the theorem extensively.

Corollary 3.1 (Poincaré–Hopf theorem for continuous vector fields) *Let* M *be as in Theorem 3.6. Then, for any continuous vector field $X \in \mathfrak{X}^{r \geq 0}(M)$ with only isolated zeroes $\{p_i^*\}_{i \in \mathcal{I}} \subset$ M, (3.14) holds.*

Proof (sketch) As illustrated on [34, p. 23], Theorem 3.3 asserts the existence of a smooth map $Y : M \to T$M being δ-close and homotopic to X. We cannot simply assume that this map will also be a vector field. However, let $\varphi = \pi \circ Y$ for the natural projection $\pi : T$M $\to$ M, which we can assume to be a diffeomorphism. Now let $X^\infty = Y \circ \varphi^{-1} : M \to T$M and see that $\pi \circ X^\infty = \mathrm{id}_M$. Then the result follows from the homotopy invariance of I, i.e., $I(X, Z_\pi(M)) = I(X^\infty, Z_\pi(M)) = \chi(M)$.

For linear systems $\dot{x}(t) = Ax(t)$, with $\det(A) > 0$, one has $\mathrm{ind}_0(Ax) = \mathrm{sgn}\det(A)$ [22, Theorem 6.1]. This can be extended to nonlinear systems by appealing to the Hartman-Grobman theorem [22, Theorem 6.3]. One show this by showing that orientation preserving diffeomorphisms are homotopic to the identity map cf. [29], see also Example 3.3 below. The take-away is that hyperbolic (structurally stable) equilibrium points have index ±1.

Next we provide a typical non-trivial example with a degenerate differential at 0. Therefore, one cannot appeal to the hyperbolic formula from above.

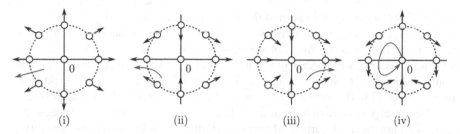

Fig. 3.4 Vector field indices with the gray lines being typical integral curves; (i): unstable radial vector field with $\text{ind}_0(X) = 1$; (ii): reflection map with $\text{ind}_0(X) = -1$; (iii): the vector field from Example 3.3 with $\text{ind}_0(X) = 0$; (iv): a vector field with homoclinic orbits and $\text{ind}_0(X) = 2$

Example 3.3 (*Degenerate equilibria*) Consider a vector field $X \in \mathfrak{X}^\infty(\mathbb{R}^2)$ given by $X(p) = (p_1^2, -p_2)$. Clearly, $p^\star = 0$ is the only equilibrium point. To compute $\text{ind}_0(X)$, see from Fig. 3.4(iii) that $y = (1, 1)$ is regular value of the map $v(X(p)) = X(p)/\|X(p)\|_2$ for both $p = 2^{-1/2}(-1, -1)$ and $p = 2^{-1/2}(1, -1)$. Hence, from (3.4), we get $\text{ind}_0(X) = 1 - 1 = 0$.

On $\mathbb{R}^2$, the vector field index corresponds to the so-called **winding number**,[7] also called the *Cauchy index* or *Poincaré index* [11] In particular, one computes for $X(p) = (p_1^2, -p_2) = (X_1, X_2)$, as in Example 3.3, the index of 0 as,

$$\text{ind}_0(X) = \frac{1}{2\pi} \int_{\mathbb{S}^1} \frac{X_1 dX_2 - X_2 dX_1}{X_1^2 + X_2^2} = \frac{1}{2\pi} \int_0^{2\pi} \frac{-\cos(\theta)^3 - 2\sin(\theta)^2 \cos(\theta)}{\cos(\theta)^4 + \sin(\theta)^2} d\theta = 0.$$

So, in dimension 2, the index corresponds to how often the vector $X_{\gamma(t)}$ rotates counter-clockwise when moving along a path $\gamma(t)$ counter-clockwise (in line with the standard orientation on $\mathbb{R}^2$) around the isolated equilibrium point [16, p. 192], see Fig. 3.4 for some more examples. See [11, Sect. 1.2] for more on the relation between the degree as defined via intersection theory or differential forms.

Next we provide as an example the **Bobylev–Krasnosel'skiĭ theorem**—that will play a central role in the remainder of this work.

Example 3.4 (*Index of isolated locally asymptotically stable equilibrium points: the Bobylev–Krasnosel'skiĭ theorem*) Consider $X \in \mathfrak{X}^r(M^n)$ with $p^\star \in M^n$ being some isolated locally asymptotically stable equilibrium point. We cannot assume $p^\star$ to be hyperbolic and follow [23, Chap. II], [22, Theorem 52.1] [40]. Without loss of generality consider X in local coordinates and assume 0 to be an isolated asymptotically stable equilibrium point. Let $\text{cl } r\mathbb{B}^n$ be a sufficiently small closed ball around 0 (occasionally written using $\mathbb{D}^n = \text{cl } \mathbb{B}^n$) such that X has no other zeroes

---

[7] See [11] for the definition using differential forms, which is outside the scope of this work. On $\mathbb{R}^2$ one can simply understand this as integrating over $d\theta$ for $\theta = \arctan(X_2/X_1)$. For instance, let $X(p) = (p_1, p_2)$, then, $\text{ind}_0(X) = (2\pi)^{-1} \int_{\mathbb{S}^1} d\theta = (2\pi)^{-1} \int_{\mathbb{S}^1} p_1 dp_2 - p_2 dp_1 = (2\pi)^{-1} \int_0^{2\pi} \cos(\theta)^2 + \sin(\theta)^2 d\theta = 1$.

on $\mathrm{cl}\, r\mathbb{B}^n$. To aid the computation of the index, we recall that the degree is invariant under homotopy. A similar notion holds for the index. We say that two vector fields, as seen as maps, are **vector field homotopic** when the entire homotopy itself does not vanish.[8] This means the vector fields themselves must be nondegenerate over their domain. Akin to Example 3.2, one can show that if this is true, the corresponding indices agree [23, Theorem 5.5]. For example, consider in Fig. 3.4 scenario (i) and (ii), the corresponding maps, from $\mathbb{S}^1$ to itself, are not homotopic and indeed, the indices do not agree. A particularly useful ramification is the following, given two nondegenerate vector fields X_1 and X_2 over W. If X_1 and X_2 are never *oppositely directed*, that is,

$$\frac{X_1(w)}{\|X_1(w)\|_2} \neq -\frac{X_2(w)}{\|X_2(w)\|_2} \quad \forall w \in W,$$

then as in Example 3.2, a convex combination of X_1 and X_2 entails a vector field homotopy, see also [23, Theorem 5.6]. Now consider the vector field $-X$ and its relation to the flow φ_X^t

$$\lim_{t \downarrow 0} \frac{p - \varphi_X^t(p)}{t} = -X(p). \tag{3.15}$$

By continuity and the fact that t is nonnegative, (3.15) implies that $p - \varphi_X^t(p)$ and $-X(p)$ will not be of opposite sign for sufficiently small $t > 0$. However, by asymptotic stability $p - \varphi_X^t(p) \neq 0$ for $t > 0$. Hence, we have constructed a vector field homotopy. However, asymptotic stability also implies that for $t \to +\infty$ the map $p \mapsto p - \varphi^t(p)$ eventually tends to the identity map. This proves that $\mathrm{ind}_0(-X) = \mathrm{ind}_0(\mathrm{id}_{\mathrm{cl}\, r\mathbb{B}^n}) = 1$ cf. (3.12). For $\mathrm{ind}_0(X)$, observe that for a map g over some n-dimensional domain $\deg(g) = (-1)^n \deg(-g)$ such that $\mathrm{ind}_0(X) = (-1)^n$. Then, by the invariance under diffeomorphisms (3.13) we get that $\mathrm{ind}_{p^\star}(X) = (-1)^n$.

The index result from Example 3.4 appeared for the first time in [4] and was largely extended and popularized by Krasnosel'skiĭ and Zabreĭko [22]. However, it is likely that the results where known, e.g., to Poincaré [35, Chap. XVIII] and Anosov [4], presumably since it appeared to be an *"obvious fact"* [4, p. 1043]. With respect to that body of literature is important to remark that the *rotation* of a vector field was the invariant of choice. For all practical purposes in this work, that concept is the same as the vector field index. For the subtle difference see [46].

Results analogous to Example 3.4 for Lyapunov stable- or attractive isolated equilibrium points are less transparent, but motivated by the work of Zabczyk [45] we provide a short remark. One severe complication is that asymptotic stability is locally of interest, just as Lyapunov stability, solely zooming in on attractivity, however, is mostly interesting on the global level due to the interplay with the (global) topology. For example, consider *globally* attractive isolated equilibrium points on $\mathbb{S}^1$ and $\mathbb{S}^2$. By the Poincaré–Hopf theorem, those equilibrium points must have vector field index 0 and 2, respectively. If those points would be merely *locally* attractive, the indices could be -1 and 1, respectively. As was already pointed out in [22], it is known that

---

[8] We will not appeal to *proper* homotopy theory here, or *"nonsingular deformations"*cf. [22].

equilibrium points that are merely Lyapunov stable can in general have any index [5]. For attractivity the statement is also subtle and depends on the domain over which the system is attractive. From Example 3.4 we see that locally, the arguments of local asymptotic stability carry over. However, see that in both cases we exploit the properties of a continuous flow. In [30] this is relaxed, the vector field is continuous, but solutions are not necessarily unique nor do they necessarily depend continuously on initial conditions. See also the proof of Theorem 6.2 for a way around exploiting the direct existence of a flow.

Example 3.5 (*Case study Sect.* 1.3; *Lie groups*) Consider any compact Lie group G^n with $n \geq 1$. As one can construct a non-vanishing smooth vector field on G^n by pushing-forward any fixed non-zero $v \in T_e G^n$ under some left translation L_g [27, Theorem 8.37], it follows from Theorem 3.6 that $\chi(G^n) = 0$, compactness is important here. We return to this frequently.

To make use of the Poincaré–Hopf theorem one needs to assert that an appropriate vector field exists. This can be shown using Thom's transversality theorem.

Proposition 3.5 (Existence of vector fields with isolated equilibrium points) *On every smooth compact manifold* M *there exists a vector field with only finitely many isolated zeroes.*

Recall Example 3.3, the following result due to Hopf shows that, up to homotopy, equilibrium points with vector field index 0 can be ignored. Compactness is key here.

Proposition 3.6 (Nowhere-vanishing vector fields [16, p. 146]) *A compact, connected, oriented, smooth manifold* M *has* $\chi(M) = 0$ *if and only if there exists a continuous nowhere-vanishing vector field* X *on* M.

See Example 5.1 for an application of Proposition 3.6 and see Chapter 6 and Section 8.2 for further generalizations of Theorem 3.6.

For references on differential topology, see for example [6, 13, 16, 18, 26, 27, 29]. For more on degree theory in particular, see [11, 16, 18, 33, 46] and [9, 31] in the context of control theory. See [14] for an exposition on the generality of index theory and [15] for a general, beyond continuity, axiomatic treatment of index theory.

References

1. Altafini C (2007) Feedback stabilization of isospectral control systems on complex flag manifolds: application to quantum ensembles. IEEE T Automat Contr 52(11):2019–2028
2. Belabbas MA (2013) On global stability of planar formations. IEEE T Automat Contr 58(8):2148–2153
3. Bobylev N, Emel'yanov SV, Korovin S (2012) Geometrical methods in variational problems. Springer, Dordrecht
4. Bobylev N, Krasnosel'skiĭ M (1974) Deformation of a system into an asymptotically stable system. Automat Remote Control 35(7):1041–1044

5. Bonatti C, Villadelprat J (2002) The index of stable critical points. Topol Appl 126(1):263–271
6. Bredon GE (1993) Topol Geometry. Springer, New York
7. Byrnes CI (2008) On Brockett's necessary condition for stabilizability and the topology of Liapunov functions on $\mathbb{R}^n$. Commun Inf Syst 8(4):333–352
8. Byrnes CI, Isidori A (1989) New results and examples in nonlinear feedback stabilization. Syst Control Lett 12(5):437–442
9. Coron J-M (2007) Control and nonlinearity. American Mathematical Society, Providence
10. Demichelis S, Ritzberger K (2003) From evolutionary to strategic stability. J Econ Theor 113(1):1–25
11. Dinca G, Mawhin J (2021) Brouwer Degree. Birkhäuser, Cham
12. Farber M (2004) Instabilities of robot motion. Topol Appl 140(2–3):245–266
13. Golubitsky M, Guillemin V (1973) Stable mappings and their singularities. Springer, New York
14. Gottlieb DH (1990) Vector fields and classical theorems of topology. Rend Sem Mat Fis Mil 60(1):193–203
15. Gottlieb DH, Samaranayake G (1995) The index of discontinuous vector fields. N Y J Math 1:130–148
16. Guillemin V, Pollack A (2010) Differential topology. American Mathematical Society, Providence
17. Hatcher A (2002) Algebraic topology. Cambridge University Press, Cambridge
18. Hirsch MW (1976) Differential topology. Springer, New York
19. Hopf H (1989) Differential geometry in the large. Springer, Berlin
20. Kalabić UV, Gupta R, Di Cairano S, Bloch AM, Kolmanovsky IV (2017) MPC on manifolds with an application to the control of spacecraft attitude on SO(3). Automatica 76:293–300
21. Koditschek DE, Rimon E (1990) Robot navigation functions on manifolds with boundary. Adv Appl Math 11(4):412–442
22. Krasnosel'skiǐ A, Zabreiko PP (1984) Geometrical methods of nonlinear analysis. Springer, Berlin
23. Krasnosel'skiǐ MA (1968) The operator of translation along the trajectories of differential equations. American Mathematical Society, Providence
24. Kvalheim MD, Gustafson P, Koditschek DE (2021) Conley's fundamental theorem for a class of hybrid systems. SIAM J Appl Dyn Syst 20(2):784–825
25. Kvalheim MD, Koditschek DE (2002) Necessary conditions for feedback stabilization and safety. J Geom Mech
26. Lee JM (2011) Introduction to topological manifolds. Springer, New York
27. Lee JM (2012) Introduction to smooth manifolds. Springer, New York
28. McLennan A (2018) Advanced fixed point theory for economics. Springer, Singapore
29. Milnor J (1965) Topology from the differentiable viewpoint. Princeton University Press, Princeton
30. Orsi R, Praly L, Mareels I (2003) Necessary conditions for stability and attractivity of continuous systems. Int J Control 76(11):1070–1077
31. Ortega R (1995) Some applications of the topological degree to stability theory. In: Topological methods in differential equations and inclusions. Springer, Dordrecht, pp 377–409
32. Ortega R (1996) A criterion for asymptotic stability based on topological degree. In: Proceedings of the first world congress of nonlinear analysts, pp 383–394
33. Outerelo E, Ruiz JM (2009) Mapping degree theory. American Mathematical Society, Providence
34. Palis JJ, De Melo W (1982) Geometric theory of dynamical systems: an introduction. Springer, New York
35. Poincaré H (1886) Sur les courbes définies par les équations différentielles (iv). J Math Pures Appl 4(2):151–217
36. Rijk FJ, Vorst AC (1983) Equilibrium points in an urban retail model and their connection with dynamical systems. Reg Sci Urban Econ 13(3):383–399
37. Sánchez-Gabites J (2010) Unstable attractors in manifolds. T Am Math Soc 362(7):3563–3589

38. Simsek A, Ozdaglar A, Acemoglu D (2007) Generalized Poincare-Hopf theorem for compact nonsmooth regions. Math Oper Res 32(1):193–214
39. Steenrod N (1951) The topology of fibre bundles. Princeton University Press, Princeton
40. Thews K (1989) Der abbildungsgrad von vektorfeldern zu stabilen ruhelagen. Arch Math 52(1):71–74
41. Tu LW (2007) An introduction to manifolds. Springer, New York
42. Vakhrameev S (1994) Morse theory and Lyusternik-Shnirelman theory in geometric control theory. J Math Sci 71(3):2434–2485
43. Varian HR (1975) A third remark on the number of equilibria of an economy. Econometrica 43(5/6):985–986
44. Ye M, Liu J, Anderson BD, Cao M (2021) Applications of the Poincare-Hopf theorem: epidemic models and Lotka-Volterra systems. IEEE T Automat Contr
45. Zabczyk J (1989) Some comments on stabilizability. Appl Math Opt 19(1):1–9
46. Zabrejko P (1997) Rotation of vector fields: definition, basic properties, and calculation. In: Topological nonlinear analysis II. Birkhäuser, Boston, pp 445–601

Chapter 4
Algebraic Topology

4.1 Singular Homology

First, we briefly introduce homology groups, for a complete—or even axiomatic—treatment, see for example [1, 4, 8]. In particular, there is a multitude of homology theories, all with their relative merits. We highlight Eilenberg's *singular* homology. The intuition goes back to Riemann and Poincaré and is as follows. Compact two dimensional surfaces can be characterized by their *genus*, that is, the number of holes (or handles). Extending this, one can consider classifying topological spaces based on how many k-dimensional "*holes*" they have and so forth. Here, the dimension of the hole should be understood as a the smallest dimension of a closed manifold enclosing the hole, e.g., $\mathbb{S}^1$ is said to have a single 1-dimensional hole.

More formally, let X be a topological space and let $\sigma : \Delta^k \to \mathsf{X}$ be any continuous map from the ***standard k-simplex*** $\Delta^k = \{p \in \mathbb{R}^{k+1} : \sum_{i=0}^k p_i = 1, \ p_i \geq 0 \text{ for } i = 0, 1, \ldots, k\}$ into X, called a ***singular k-simplex***. Now recall the introductory remarks on *groups* from Sect. 1.3, yet, to work with these singular maps we need to introduce more concepts from algebra. Let G be an *Abelian* (commutative) group, then $\mathscr{B} \subseteq \mathsf{G}$ is a *basis* of G when G is the smallest subgroup of G that contains $\mathscr{B}$ and is such that every $g \in \mathsf{G}$ can be expressed as a formal sum (meaning, for general "$+$") $g = \sum_i \alpha_i b_i$ with $\alpha_i \in \mathbb{Z}$, $b_i \in \mathscr{B}$ and only finitely many α_i being non-zero [5, Theorem I.2.8]. Then, while omitting the (motivating) details, given any set Y the so-called *free Abelian group* generated by Y is given by $\{\sum_i \alpha_i y_i : y_i \in Y, \ \alpha_i \in \mathbb{Z}$ and finitely many α_i are non-zero $\}$ [5, Chap. II], which is an additive group.

We also recall the notion of a ***group homomorphism***, that is, given two groups $(\mathsf{G}, \cdot^{(\mathsf{G})})$, $(\mathsf{H}, \cdot^{(\mathsf{H})})$, a map $z : \mathsf{G} \to \mathsf{H}$ such that $z(g_1 \cdot^{(\mathsf{G})} g_2) = z(g_1) \cdot^{(\mathsf{H})} z(g_2)$ for all $g_1, g_2 \in \mathsf{G}$. The set of group homomorphisms between G and H is denoted by $\mathrm{Hom}(\mathsf{G}, \mathsf{H})$. In case H is Abelian, $(\mathrm{Hom}(\mathsf{G}, \mathsf{H}), \cdot)$ is itself an Abelian group with the group operation being defined pointwise by $(z_1 \cdot z_2)(g) = z_1(g) \cdot^{(\mathsf{H})} z_2(g)$ for all $z_1, z_2 \in \mathrm{Hom}(\mathsf{G}, \mathsf{H})$ and all $g \in \mathsf{G}$.

Now let $C_k(\mathsf{X})$ denote the free Abelian group generated by all singular k-simplices, called the ***singular chain group***, containing elements, called k-***chains***, of the form $\sum_i \alpha_i \sigma_i$ for σ_i a singular k-simplex. For instance, for $C_1(\mathsf{X})$, one can think of $\sigma_i : [0, 1] \to \mathsf{X}$ as giving rise to a path (or point) in X. Recall that paths enclosing on itself

© The Author(s) 2023
W. Jongeneel and E. Moulay, *Topological Obstructions to Stability and Stabilization*,
SpringerBriefs in Control, Automation and Robotics,
https://doi.org/10.1007/978-3-031-30133-9_4

(loops), and in general maps from n-spheres, contain a lot of topological information. However, this *homotopy* approach does not necessarily detect all *"holes"* we are after in an intuitive manner, e.g., Hopf found a non-trivial map from the 3-sphere to the 2-sphere. This is one reason to use the singular chain groups instead. In particular, one can define a *boundary operator* ∂_k that acts on $C_k(\mathsf{X})$ as

$$\cdots \overset{\partial_{k+1}}{\to} C_k(\mathsf{X}) \overset{\partial_k}{\to} C_{k-1}(\mathsf{X}) \overset{\partial_{k-1}}{\to} \cdots 0 \tag{4.1}$$

and satisfies $\partial_k \circ \partial_{k+1} = 0$ (boundaries have no boundaries). In the context of singular homology this map can be made explicit. To that end, define the *face embedding* $F_{i,k} : \Delta^{k-1} \hookrightarrow \Delta^k$ as follows. Let $\{e_0, e_1, \ldots, e_\ell\}$ be the set of vertices of Δ^ℓ. Then, $F_{i,k}$ is such that it maps Δ^{k-1} to the face opposite to the vertex $e_i \in \Delta^k$. Note that Δ^{k-1} itself is a face of Δ^k. Now for $\sigma \in C_k(\mathsf{X})$ the boundary operator can be defined as $\partial_k \sigma = \sum_{i=0}^{k} (-1)^i \sigma \circ F_{i,k}$, where $\partial_k \circ \partial_{k+1} = 0$ can be verified.

Then, a k-chain $c \in C_k(\mathsf{X})$ is called a k-*cycle* when $\partial_k c = 0$. Differently put, the set $\ker(\partial_k) \subseteq C_k(\mathsf{X})$ contains all k-cycles. See also that when $b \in C_{k+1}(\mathsf{X})$, then, $\partial_{k+1} b \in \ker(\partial_k)$ since $\partial_k(\partial_{k+1} b) = 0$. Now, the kth *singular homology group* of X is defined as $H_k(\mathsf{X}; \mathbb{Z}) = H_k(\mathsf{X}) = \ker(\partial_k)/\mathrm{im}(\partial_{k+1})$. As such, $H_k(\mathsf{X}; \mathbb{Z}) = 0$ when all k-cycles are boundaries of $(k+1)$-chains, that is, there are no k-dimensional holes.

Example 4.1 (*The* 0*th singular homology group*) Let X be path-connected.[1] we will follow [6] and show that $H_0(\mathsf{X}; \mathbb{Z}) = H_0(\mathsf{X}) = \ker(\partial_0)/\mathrm{im}(\partial_1) \simeq \mathbb{Z}$. First of all, by (4.1) see that $\ker(\partial_0) = C_0(\mathsf{X})$. Now for any 0-chain $c = \sum_i \alpha_i \sigma_i$ construct the *index map* $I : C_0(\mathsf{X}) \to \mathbb{Z}$ by $I(c) = \sum_i \alpha_i$. Evidently, this map is a surjective homomorphism. We will show that $\ker(I) = \mathrm{im}(\partial_1)$, which implies that $\mathbb{Z}$ is isomorphic to $H_0(\mathsf{X})$ by the *first isomorphism theorem* for groups.[2] Pick any singular 1-simplex σ, then $\partial \sigma = \sigma(1) - \sigma(0)$ and indeed $I(\partial \sigma) = 0$. This implies that $\mathrm{im}(\partial_1) \subseteq \ker(I)$. For the other direction, fix a point in X, say x', and let $\psi(x)$ denote a continuous curve from $x \in \mathsf{X}$ to x', which always exists as X is path-connected. This means that for any 0-chain $c = \sum_i \alpha_i x_i$ (recall that $\sigma_i = x_i$ in this case), we have $\partial(\alpha_i \sum_i \psi(x_i)) = \sum_i \alpha_i x_i - I(c) x'$. Therefore, if $I(c) = 0$, then c can be written as the boundary of a 1-chain and hence $\ker(I) \subseteq \mathrm{im}(\partial_1)$. This concludes showing that $H_0(\mathsf{X}) \simeq \mathbb{Z}$. When X consists out of p components, this argument is generalized to showing that $H_0(\mathsf{X}) \simeq \mathbb{Z}^p$. See also [1, p. 172] for a similar explanation.

To provide another important example, $H_k(\mathbb{S}^{n \geq 1}; \mathbb{Z}) \simeq \mathbb{Z}$ for $k \in \{0, n\}$ and 0 otherwise. Similarly, given a subspace A of X, one can consider the homology group of X *"modulo* A*"* as follows. Define the *relative* chain group $C_k(\mathsf{X}, \mathsf{A}) = C_k(\mathsf{X})/C_k(\mathsf{A})$ and analogously the (relative) boundary operator $\partial_{\mathsf{A},k} : C_k(\mathsf{X}, \mathsf{A}) \to C_{k-1}(\mathsf{X}, \mathsf{A})$. Then, the k-th *singular homology group* of X *relative* to A is defined as $H_k(\mathsf{X}, \mathsf{A}; \mathbb{Z}) = H_k(\mathsf{X}, \mathsf{A}) = \ker(\partial_{\mathsf{A},k})/\mathrm{im}(\partial_{\mathsf{A},k+1})$ [4, p. 115]. When $\mathsf{A} \neq \emptyset$ is closed

---

[1] A topological space X is said to be *path-connected* when for any $x, y \in \mathsf{X}$ there is a *continuous* function $g : [0, 1] \to \mathsf{X}$ such that $g(0) = x$ and $g(1) = y$.

[2] The statement is as follows, let G and H be groups and let z be a group homomorphism. Then, $\mathrm{im}(z)$ is isomorphic to $\mathsf{G}/\ker(z)$ [6, Theorem C.10].

and a neighbourhood deformation retract of X, then $H_k(X, A; \mathbb{Z}) = \tilde{H}_k(X/A; \mathbb{Z})$ [4, Proposition 2.22], for $\tilde{H}_k$ the *reduced homology*, i.e., $H_k \simeq \tilde{H}_k$ for $k > 0$ and $H_0 \simeq \tilde{H}_0 \oplus \mathbb{Z}$ [4, p. 110]. Whenever k is irrelevant, we write $H_{(\cdot)}$, where $H_\bullet$ is also common notation. Omitting details, when X is a *compact* manifold, then $H_k(X)$ is a finitely-generated Abelian group such that the **rank** of $H_{(\cdot)}$ is simply the number of $\mathbb{Z}$ summands used to describe the group [4, Sect. 2.2]. However, $H_k(X)$ is by no means finitely-generated in general, consider a plane with uncountably many holes.

Now, dual to the homology groups, one can define via $C^k(X) = \mathrm{Hom}(C_k(X), \mathbb{Z})$ the k-th **singular cohomology group** of X via the so-called *induced coboundary operator* $\delta^k : C^k(X) \to C^{k+1}(X)$ as $H^k(X; \mathbb{Z}) = H^k(X) = \ker(\delta^k)/\mathrm{im}(\delta^{k+1})$ [4, 6]. Then, *Poincaré duality* allows for linking homology and cohomology groups of X [4, Sect. 3.3], e.g., $H_k(M^m; \mathbb{Z}) \simeq H^{m-k}(M^m; \mathbb{Z})$, for appropriate M^m (see below).

The power of *singular* homology does not necessarily lie in computation, but in the ability to prove relationships between several homology groups. To that end, given a continuous map $G : X \to Y$ define the homomorphism $G_\# : C_k(X) \to C_k(Y)$ by $G_\#(\sigma) = G \circ \sigma$ for any singular k-simplex $\sigma \in C_k(X)$. The explicit formula for the boundary operator reveals that $G_\#(\partial_k \sigma) = \partial_k(G_\#(\sigma))$. Note that at the LHS of this equality the operator ∂_k acts on $C_k(X)$ whereas on the RHS ∂_k acts on $C_k(Y)$. This means that $G_\#$ maps cycles to cycles and so forth. As such, $G_\#$ *induces* a homomorphism $G_\star : H_k(X, \mathbb{Z}) \to H_k(Y; \mathbb{Z})$. It readily follows that for two continuous maps $G_1 : X \to Y$ and $G_2 : Y \to Z$ we have $(G_2 \circ G_1)_\star = G_{2\star} \circ G_{1\star}$. We are now equipped to state the following lemma.

Lemma 4.1 (Homology homotopy invariance [4, Corollary 2.11]) *Let $G : X \to Y$ be a homotopy equivalence, then, the induced homomorphism $G_\star : H_k(X; \mathbb{Z}) \to H_k(Y; \mathbb{Z})$ on singular homology is an isomorphism for any $k \geq 0$.*

Lemma 4.1 is commonly proved by first proving that for homotopic maps their induced homomorphims are equivalent [4, Theorem 2.10]. Then, one uses that if $G : X \to Y$ is a homotopy equivalence, there must exist a map $G' : Y \to X$ such that $G' \circ G \simeq_h \mathrm{id}_X$ and similarly, $G \circ G' \simeq_h \mathrm{id}_Y$. However, this implies that $G'_\star \circ G_\star = (\mathrm{id}_X)_\star$ and $G_\star \circ G'_\star = (\mathrm{id}_Y)_\star$. Therefore, $G_\star$ must be an *isomorphism*.

An elementary implication of Lemma 4.1—which will be of use in Chap. 6—is that by homotopy equivalence between $\mathbb{R}^n \setminus \{0\}$ and $\mathbb{S}^{n-1}$ for $n \geq 2$, the singular homology groups of $\mathbb{R}^n \setminus \{0\}$, for $n \geq 2$, become

$$H_k(\mathbb{R}^n \setminus \{0\}; \mathbb{Z}) \simeq \begin{cases} \mathbb{Z} & \text{if } k \in \{0, n-1\} \\ 0 & \text{otherwise} \end{cases}. \tag{4.2}$$

Now, let $G : \mathbb{S}^n \to \mathbb{S}^n$ be a continuous map, then, the degree of G as described in Sect. 3.4 can be equivalently defined by means of the induced homomorphism $G_\star$, that is, as the integer $\deg(G)$ such that $G_\star(H_n(\mathbb{S}^n; \mathbb{Z})) = \deg(G)H_n(\mathbb{S}^n; \mathbb{Z})$, e.g., see [1, Chap. IV]. This definition of the degree also immediately works for maps $G : M^m \to N^n$ over oriented, connected, compact manifolds as $H_n(M^n; \mathbb{Z}) \simeq \mathbb{Z}$ (by

Fig. 4.1 Vector field on a
triangulation

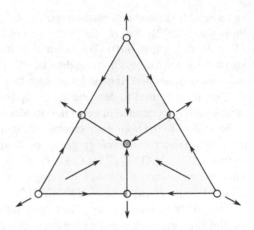

the *universal coefficient theorem* [1, Chap. V]) [7, Sect. III.2]. We return to this
viewpoint below and in Chap. 6.

4.2 The Euler Characteristic

To link the previous chapter with the geometric definition of the Euler characteristic
as set forth in Sect. 2.3, find Fig. 4.1. A single triangle is constructed using 3 vertices,
3 edges and 1 face, as such this adds up to an Euler characteristic of 1. Similarly, see
that we can construct a vector field with 3 sources, 3 saddles and 1 sink, adding up
the vector field indices agrees with the Euler characteristic.

This can be formalized using homology theory, which will be briefly outlined
below. First, one generalizes the 2-dimensional Euler characteristic formula by
appealing to Whitehead's CW complexes [4, 6]. Informally put, a CW complex
is a space constructed by glueing together k-cells, that is, topological k-dimensional
balls. These cells can be of different dimension and the glue is applied to the boundary of the cells. More formally, let X_0 be some discrete space and construct X_1 by
attaching some collection $\{C_j\}_{j \in J}$ of open 1-cells to X_0 via a collection of continuous
maps $\varphi_j : \partial C_j \to X_0$, that is, $X_1 = X_0 \cup \varphi(\sqcup_j \partial C_j)$. One can continue this proce-
dure and construct $X_2 \subseteq X_3 \subseteq \cdots \subseteq X_n$ by attaching open k-cells of appropriate
dimension. If a topological space X can be written as X_n for some $n \geq 0$, as defined
above, then X has a cell decomposition $\mathscr{C}$, where the 0-cells are given by X_0, the
1-cells by $X_1 \setminus X_0$ and so forth. The pair $(\mathsf{X}, \mathscr{C})$ is called a *cell complex*. This cell
complex is a **CW complex** when it satisfies the following two properties

1. The closure of each cell is contained in a union of finitely many cells;
2. The topology of X is coherent with $\{\operatorname{cl} C : C \in \mathscr{C}\}$.

See [6, p. 135] for examples that fail to meet both conditions.

Definition 4.1 (*The (combinatorial) Euler characteristic* [6, Chap. 6]) Let X^n be a finite dimensional CW complex, with n_k the number of k-cells of X^n. Then, the Euler characteristic of X^n is defined as

$$\chi(X^n) = \sum_{k=0}^{n} (-1)^k n_k. \tag{4.3}$$

Although all equivalent under compactness assumptions, e.g., see Theorem 4.2 below, we call (4.3) the *combinatorial* definition of the Euler characteristic. This definition shows in particular that for non-compact spaces homotopy invariance of $\chi(\cdot)$ does not necessarily hold, e.g., compare χ for an open and a closed interval.

Example 4.2 (*The CW structure is not unique: the 2-sphere*) Recall that the sphere $\mathbb{S}^2$ can be constructed from a 2-cell (a disk) and 0-cell (a point), hence $\chi(\mathbb{S}^2) = 2$. Similarly, one could fix two poles and construct $\mathbb{S}^2$ from two 0-cells, two 1-cells (two intervals) and two 2-cells, adding up to $\chi(\mathbb{S}^2) = 2$.

See also [9] for a less straightforward matrix manifold example. The combinatorial formula (4.3) is particularly appealing due to the following result.

Proposition 4.1 (CW equivalences [4, Corollary A.12]) *Any compact topological manifold is homotopy equivalent to a finite CW complex.*

Next, we provide the homological definition of $\chi(X^n)$, here, $b_k = \operatorname{rank} H_k(X^n; \mathbb{Z})$ denotes the k-th **Betti number** of X^n, i.e., when $H_k(X; \mathbb{Z})$ is finitely generated, we have $H_k(X^n; \mathbb{Z}) \simeq \mathbb{Z}^{b_k} \oplus T_k$, for T_k the *torsion* [1, p. 258, 282]. Compute for instance $H_{(\cdot)}(\mathbb{RP}^n; \mathbb{Z})$ to see $T_{(\cdot)} \neq 0$. In that sense, we should speak of *"holes" and "twists"*.

Theorem 4.1 (The (homological) Euler characteristic [4, Theorem 2.44]) *Let X^n be a finite CW complex. Then, the Euler characteristic of X^n equals*

$$\chi(X^n) = \sum_{k=0}^{n} (-1)^k b_k. \tag{4.4}$$

It follows from *Poincaré duality* that for closed[3] oriented manifolds one has $\operatorname{rank} H_k(M^m; \mathbb{Z}) = \operatorname{rank} H^{m-k}(M^m; \mathbb{Z})$, and by *the universal coefficient theorem* that $\operatorname{rank} H_k(M^m; \mathbb{Z}) = \operatorname{rank} H^k(M^m; \mathbb{Z})$ e.g., see [4, Corollary 3.37]. However, then we have

$$\operatorname{rank} H_{m-k}(M^m; \mathbb{Z}) = \operatorname{rank} H_k(M^m; \mathbb{Z}), \tag{4.5}$$

see [4, Sect. 3.3] for the omitted details. A similar argument holds for non-orientable manifolds, such that by (4.4) the following result follows.

Corollary 4.1 (Odd-dimension Euler characteristic [4, Corollary 3.37]) *A closed manifold M of odd-dimension has $\chi(M) = 0$.*

---

[3] In this context, **closed** means, traditionally, compact and without boundary.

Proof (*Sketch*) Combining (4.4) with (4.5): $\chi(M) = b_0 - b_1 + \cdots + b_1 - b_0 = 0$.

Note that for oriented manifolds, Corollary 4.1 can also be shown using oriented intersection theory, cf. (3.5). When M has a boundary, Corollary 4.1 is not true, consider the interval $[0, 1]$. We will return to Corollary 4.1 frequently.

We end with one of the pillars of topology, linking the seemingly different definitions of the Euler characteristic. See Sect. 8.3 for comments on Morse theory.

Theorem 4.2 (The Euler characteristic [2, Theorem 8.6.6]) *Let M be a closed, oriented, smooth manifold, then the corresponding combinatorial Euler characteristic (4.3), homological Euler characteristic (4.4) and the Euler characteristic from oriented intersection theory (3.6) all agree.*

Proof (*Sketch*) Recall the relation between Morse indices and cells in a CW structure, e.g., see Theorem 8.2 below. Then, following [2], construct a Morse function $g : M^m \to \mathbb{R}$ with m_k critical points of index k, for $k = 1, \ldots, m$. Now, using Theorem 8.2 and Definition 4.1 it follows that $\chi(M^m) = \sum_k (-1)^k m_k$. However, note that the vector field index for an equilibrium point (critical point of g) with Morse index k equals $(-1)^k$, and we have m_k of them. As such, Theorem 3.6 (Poincaré–Hopf) tells us that $\chi(M^m) = \sum_k (-1)^k m_k$, which is exactly what we had before. Then, the relation to homology follows from Theorem 4.1.

Observe that compactness is important as $\chi(\mathbb{R}^n) = (-1)^n$ according to the combinatorial definition (4.3) while $\chi(\mathbb{R}^n) = 1$ according to singular homology (4.4).

We refer the reader to [3] for an illustrated introduction to algebraic topology and to [1, 4, 8] for complete treatments.

References

1. Bredon GE (1993) Topology and geometry. Springer, New York
2. Burns K, Gidea M (2005) Differential geometry and topology: with a view to dynamical systems. Chapman and Hall/CRC, Boca Raton
3. Ghrist RW (2014) Elementary applied topology. Createspace, Seattle
4. Hatcher A (2002) Algebraic topology. Cambridge University Press, Cambridge
5. Hungerford TW (1974) Algebra. Springer, New York
6. Lee JM (2011) Introduction to topological manifolds. Springer, New York
7. Outerelo E, Ruiz JM (2009) Mapping degree theory. American Mathematical Society, Providence
8. Spanier EH (1981) Algebraic topology. McGraw-Hill, New York
9. Tomei C (1984) The topology of isospectral manifolds of tridiagonal matrices. Duke Math J 51(4):981–996

Chapter 5
Dynamical Control Systems

5.1 Dynamical Systems

We start by clarifying notation and terminology. Let M be a topological manifold, $(\mathcal{T}, \cdot)$ a commutative topological group with $e \in \mathcal{T}$ denoting its *identity* element, i.e., $s \cdot e = e \cdot s = s$ $\forall s \in \mathcal{T}$ and let $\varphi : \mathcal{T} \times M \to M$ be a continuous map. The triple $(M, \mathcal{T}, \varphi)$ is a **dynamical system** when the following axioms are satisfied

(i) the *identity property*: $\varphi(e, p) = p$ for all $p \in M$; and
(ii) the *group property*: $\varphi(s \cdot t, p) = \varphi(s, \varphi(t, p))$ for all $s, t \in \mathcal{T}$ and $p \in M$.

When $\mathcal{T} = \mathbb{R}$ the map φ is said to be a (global) *flow*. In that case, the axioms are conveniently written as $\varphi^0 = \mathrm{id}_M$ and $\varphi^s \circ \varphi^t = \varphi^{s+t}$ $\forall s, t \in \mathbb{R}$, with φ^t denoting the homeomorphism $\varphi(t, \cdot) : M \to M$. Indeed, see that a flow naturally induces the homotopy equivalence $\varphi^t(U) \simeq_h \varphi^s(U)$ for any $s, t \in \mathbb{R}$ and open set $U \subseteq M$. See also that $(M, \mathcal{T}, \varphi)$ is *time-invariant* in the sense that $\varphi(s, p)$ only depends on the current point $p \in M$ and the "*time*" s to *propagate*, i.e., $\varphi(s, \varphi(s^{-1}, \varphi(s, p))) = \varphi(s, p)$.

Let M be a smooth manifold and let $X \in \mathfrak{X}^r(M)$, with $r \in \mathbb{N} \cup \{\infty\}$, denote a C^r-smooth **vector field** over M, that is, $X : M \to TM$ is a C^r-smooth map and $\pi_p \circ X = \mathrm{id}_M$ for π_p being the canonical projection $\pi_p : TM \to M$ defined by $\pi_p : (p, v) \mapsto p$, sometimes simply written as π. Equivalently, C^r vector fields X on M can be understood as C^r sections of the tangent bundle, denoted $X \in \Gamma^r(TM)$. The evaluation of $X \in \mathfrak{X}^r(M)$ at $p \in M$ is a tangent vector $X(p) = X_p \in T_pM$.

Then, a differentiable curve $\xi : \mathcal{I} \subseteq \mathbb{R} \to M$, for some appropriate interval $\mathcal{I}$, is called an **integral curve** of the vector field X when $\dot{\xi}(t) = X(\xi(t))$ for all $t \in \mathcal{I}$. A manifestation of a flow that will be of interest is as a map parametrizing all integral curves of a vector field. Given a vector field $X \in \mathfrak{X}^r(M)$, then via the relation

$$\frac{\mathrm{d}}{\mathrm{d}t}\varphi^t(p)\bigg|_{t=t'} = X(\varphi^{t'}(p)), \tag{5.1}$$

we can define a C^r local flow $\varphi : \mathrm{dom}(\varphi) \to M$, where the regularity is commonly with respect to the second argument of the flow. See that the map $t \mapsto \varphi^t(p)$ is always at least C^1 when the flow is generated by a continuous vector field. A flow

© The Author(s) 2023
W. Jongeneel and E. Moulay, *Topological Obstructions to Stability and Stabilization*,
SpringerBriefs in Control, Automation and Robotics,
https://doi.org/10.1007/978-3-031-30133-9_5

is not necessarily well-defined for all $t \in \mathbb{R}$ or even all $t \geq 0$,[1] e.g., consider $\dot{x} = x^2$ with $x(0) = 1$. When X *does* give rise to a globally well-defined flow the vector field is said to be **complete**. Regarding the case study from Sect. 1.3, every left-invariant vector field on a Lie group is complete [43, Theorem 9.18]. In particular, any smooth vector field over a compact manifold is complete [43, Theorem 9.16]. When a flow originates from a vector field X, we denote this by φ_X.

We will follow the terminology from Chap. 3; given a smooth map G, we speak of the *differential of G* or the *pushforward under G*, denoted DG and G_*, respectively. Now we formalize what this means in case G acts on vector field or is a vector field itself. Following [43], given a smooth map $G : M \to N$ the pushforward of G at $p \in M$, that is, $DG_p : T_pM \to T_{G(p)}N$, is defined pointwise by $DG_p(X_p)g = X_p(g \circ G)$ for any smooth function $g : N \to \mathbb{R}$ and any $(p, X_p) \in TM$. Here, the action of a tangent vector X_p on a C^1 function g should be understood, in coordinates, as the directional derivative of g in the direction of X_p. To turn this pointwise pushforward, that is, all the vectors $DG_p(X_p)$, into a vector field on N, we must be able to supply any point $q \in N$ and get an element in T_qN, that is, we need p such that $G(p) = q$. Hence, we assume G to be a C^{r+1} diffeomorphism and compose with G^{-1}, this defines the pushforward G_*X of a vector field $X \in \mathfrak{X}^r(M)$ under G as the vector field $G_*X \in \mathfrak{X}^r(N)$ satisfying $(DG(X)g) \circ G^{-1} = G_*Xg$ for any smooth $g : N \to \mathbb{R}$ cf. Sect. 3.5, also understood via the following diagram

$$
\begin{array}{ccc}
M & \xrightarrow{\ G\ } & N \\
\downarrow{\scriptstyle X} & & \downarrow{\scriptstyle G_*X} \\
TM & \xrightarrow{\ DG\ } & TN
\end{array}
$$

Moreover, we make frequent use of the differential *of $X \in \mathfrak{X}^{r \geq 1}(M)$ at $p \in M$* such that $X(p) = 0$, written as $DX_p : T_pM \to T_{X(p)}TM$. With some abuse of notation this map is, however, commonly written as $DX_p : T_pM \to T_pM$ such that, in coordinates, one can consider the eigenvalues of DX_p. This simplification hinges on the fact that for $p \in M$, $X : p \mapsto (p, v)$ with $v \in T_pM$ such that the first part of $DX_p : T_pM \to T_{X(p)}TM$ is just the identity map on T_pM, hence the simplification. Note, without imposing further assumptions, this *"connection-free"* construction of DX_p will only make sense when $X_p = 0 \in T_pM$. To see this, recall that $X \in \mathfrak{X}^{r \geq 1}(M)$ induces a $C^{r \geq 1}$ (local) flow $\varphi^t : M \to M$ [43, Theorem D.5]. Now, consider the time-derivative of $D\varphi_p^t : T_pM \to T_{\varphi^t(p)}M$ and see that we work within the same tangent space when p is a fixed point of φ^t. In that case we do not need additional structure and can define $DX_p : T_pM \to T_pM$ via

$$
\frac{\mathrm{d}}{\mathrm{d}t} D\varphi_p^t v = DX_p v, \qquad v \in T_pM.
$$

---

[1] A flow defined over $\mathbb{R}_{\geq 0}$ is usually called a *semiflow*, as it does not satisfy flow property (ii).

Via (5.1) one can observe that in coordinates this construction entails the Jacobian of X. One can also see this from $T_{X(p)=0}T\mathsf{M} \simeq T_p\mathsf{M} \oplus T_p\mathsf{M}$ [2, p. 72] or by observing that the non-Euclidean part of the covariant derivative vanishes [42, Chap. 4].

As most of this work will be centred around providing necessary conditions, we will not be further concerned with integrability. We point the reader to, [43, 57, Appendix D], [62, Appendix C] and note that a uniquely integrable continuous vector field $X \in \mathfrak{X}^{r\geq0}(\mathsf{M})$ under *locally Lipschitz* regularity conditions gives rise to a unique maximal flow $\varphi_X : \mathcal{I} \times \mathsf{M} \to \mathsf{M}$, for some $\mathcal{I} \subseteq \mathbb{R}$. To simplify the overall exposition, we will assume—unless stated otherwise—the following throughout.

Assumption 5.1 (*Unique global integrability*) Every vector field $X \in \mathfrak{X}^{r\geq0}(\mathsf{M})$ considered in this work is complete and uniquely integrable, that is, X gives rise to a unique global continuous flow $\varphi_X : \mathbb{R} \times \mathsf{M} \to \mathsf{M}$.

Nevertheless, notable results that can do with weaker assumptions will be highlighted. Next we introduce a notion due to Birkhoff [10, p. 197].

Definition 5.1 (*The ω-limit set* [56, p. 148]) Given a flow φ on a topological manifold M, the ω-*limit* set of $p \in \mathsf{M}$ is

$$\omega(\varphi, p) = \bigcap_{T\geq0} \mathrm{cl} \bigcup_{t\geq T}\{\varphi^t(p)\}. \tag{5.2}$$

Differently put, as we work with global flows (complete vector fields), $y \in \omega(\varphi, p)$ when there is a monotonically increasing sequence $\{t_n\}_{n\in\mathbb{N}} \subset \mathbb{R}$ with $\lim_{n\to\infty} t_n = +\infty$ such that $\lim_{n\to\infty} \varphi^{t_n}(p) = y$ [2, Proposition 6.1.2]. The ω-limit set captures any type of asymptotic recurrent behaviour, like the convergence to equilibrium points but also limit cycles and so forth, as further detailed below. Analogously, one can define the α-*limit* by reversing time. When generalizing Definition 5.1 to sets $P \subseteq \mathsf{M}$ it is important to see that $\omega(\varphi, P)$ is not necessarily equivalent to $\bigcup_{p\in P} \omega(\varphi, p)$.

Recall, given a vector field $X \in \mathfrak{X}^r(\mathsf{M})$, we call $p^\star \in \mathsf{M}$ an *equilibrium point* of X when $X_{p^\star} = 0 \in T_{p^\star}\mathsf{M}$, equivalently, $\omega(\varphi_X, p^\star) = \{p^\star\}$.[2] The point $p^\star$ is called *isolated* when there is an open neighbourhood $U \subseteq \mathsf{M}$ of $p^\star$ such that for all $p \in U \setminus \{p^\star\}$ one has $X_p \neq 0$, equivalently, $\omega(\varphi_X, p) \neq \{p\}$ for all $p \in U \setminus \{p^\star\}$. Note, this is different from an *isolated set* in the sense of Conley cf. Sect. 8.5. Away from equilibrium points the flow is locally a straight line [53, Theorem 2.26]. This *flow-box* theorem indicates why the behaviour around equilibrium points is interesting to study, but also that periodic orbits are inherently hard to study locally.

Definition 5.2 (*Hyperbolic equilibrium points* [55, p. 58]) Let $p^\star \in \mathsf{M}^m$ be an equilibrium point of $X \in \mathfrak{X}^{r\geq1}(\mathsf{M}^m)$. Then, $p^\star$ is a *hyperbolic equilibrium point* if $DX_{p^\star} : T_{p^\star}\mathsf{M}^m \to T_{p^\star}\mathsf{M}^m$ defines a hyperbolic linear vector field, that is, all eigenvalues (in coordinates) of $DX_{p^\star}$ in $\mathbb{R}^{m\times m}$ have a non-zero real part.

---

[2] Note, here we exploit Assumption 5.1, if the integral curves would not be unique then zeroing the vector field is not sufficient, e.g., consider $\dot{x} = \sqrt{|x|}$.

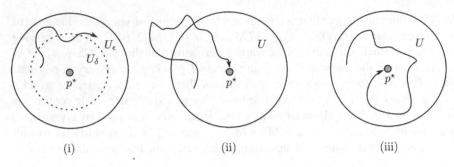

Fig. 5.1 Definition 5.3: (i) Lyapunov stability; (ii) attractivity; and (iii) asymptotic stability

Hyperbolic equilibrium points are generic [55, p. 58] and isolated, cf. Proposition 3.5. Next, we define the qualitative behaviour of interest for time-invariant dynamical systems.

Definition 5.3 (*Time-invariant stability notions of equilibrium points*) Given a vector field $X \in \mathfrak{X}^r(\mathsf{M})$ defining the system (5.1) we distinguish the following notions of stability for an equilibrium point $p^\star \in \mathsf{M}$ of X:

 (i) **Lyapunov stability**: for each neighbourhood U_ϵ of $p^\star$ there is neighbourhood U_δ of $p^\star$ such that for all $p_0 \in U_\delta$ one has $\varphi_X^t(p_0) \in U_\epsilon$ for all $t \geq 0$, that is, $\varphi_X^t(U_\delta) \subseteq U_\epsilon$ for all $t \geq 0$;
 (ii) **Local attractivity**: there is a neighbourhood U of $p^\star$ such that for all $p_0 \in U$ one has $\lim_{t \to +\infty} \varphi_X^t(p_0) = p^\star$;
 (iii) **Local asymptotic stability**: $p^\star$ satisfies both (i) and (ii).

If (ii) holds with $U = \mathsf{M}$, then the point $p^\star$ is globally attractive and similarly, if $p^\star$ is globally attractive and stable in the sense of Lyapunov, then, $p^\star$ is **globally asymptotically stable**, see Fig. 5.1. If (i) fails to hold, $p^\star$ is called **unstable**.

We will be mostly interested in studying asymptotic stability, however, not only regarding equilibrium *points*. A compact set $A \subseteq \mathsf{M}$ is called a local **attractor** of the flow φ when A is φ-invariant, that is, $\varphi^t(A) \subseteq A$ for all $t \in \mathbb{R}$, and there is an open neighbourhood $U \subseteq \mathsf{M}$ of A such that $\cap_{t \geq 0} \varphi^t(U) = A$. In fact, this is equivalent to saying that the invariant set A is locally asymptotically stable [28, Lemma 1.6].[3] Attractivity is clear, for Lyapunov stability, consider for example Fig. 1.1(vi), $\cap_{t \geq 0} \varphi^t(U)$ will correspond to a set which is both not closed and open, e.g., of the form $[a, b)$. Indeed, lacking Lyapunov stability or lacking local attractivity are not mutually exclusive notions, see also [26, Sect. 40].

As global asymptotic stability will turn out to be often impossible, we make a special distinction, we will consider **local asymptotic multistability**, e.g., all isolated equilibrium points of $X \in \mathfrak{X}^r(\mathsf{M})$ are locally asymptotically stable. The importance

---

[3] The literature does not agree on terminology here cf. [9, Definition V.1.5].

of this notion follows from the fact that having multiple attractors means that disturbances can qualitatively change the nominal behaviour, moving from one attractor
to another. Multistability appears for example in the study of laser dynamics [6]
and neural networks [14], with the importance of multistability being especially
acknowledged in biology, e.g., see [5, 41, 49].

So far, everything was qualitative, yet, when imposing a metric on M, stability
(the rate of convergence) can be quantified [12, Sect. 6.1.5]. See [25] for the relation
between asymptotic- and this quantitative notion called *exponential* stability.[4] Also,
a metric enables handling non-compact attractors, e.g., see [73, Sect. 3].

When the dynamical system is ***time-varying***, e.g., when $X : \mathcal{I} \times M \to TM$ with
$\mathcal{I} \subseteq \mathbb{R}$ is a continuous time-varying vector field, we need to generalize our stability
notions. First note that time-varying vector fields do not necessarily give rise to flows,
however, time-dependent generalizations are possible [43, Theorem 9.48]. One still
speaks of $\xi : \mathcal{I} \to M$ as an integral curve of the *time-varying* vector field X when
$\dot{\xi}(t) = X(t, \xi(t))$ for all $t \in \mathcal{I}$, however, one might do with a *weak* solution to the
differential equation. That is, one allows for ξ to be merely absolutely continuous and
to satisfy the differential equation almost everywhere, with respect to the Lebesgue
measure. In what follows, when a vector field is time-varying, we take this viewpoint,
see [1, Chap. 4], [12, Appendix A], [62, Appendix C] and [30] for more on solutions
of time-varying vector fields. In general, for a time-varying dynamical system the
$\epsilon - \delta$ definition of Lyapunov stability might depend on time, that is, δ might depend
on time. When this is not the case, we speak of ***uniform stability***, which coincides
with stability in case the vector field is time-invariant. Similarly, one can extend the
other notions of stability, see [36, Sect. 4.5], [18, Sect. 11.2] and Example 6.4 below.
In particular, see [69, Sect. 5.1] for illustrative examples of the aforementioned
stability notions due to Massera and Vidyasagar and see [66] for misconceptions
when it comes to uniform stability. In particular, a lack of uniformity can compromise
robustness, e.g., $\delta \to 0$ for $t \to +\infty$.

Now given an equilibrium point $p^\star \in M$ for some vector field $X \in \mathfrak{X}^r(M)$, it is
worthwhile to characterize the set of points that flow towards $p^\star$ under X.

Definition 5.4 (*Domain of attraction of an equilibrium point*) Let $p^\star$ be a local
asymptotically stable equilibrium point of $X \in \mathfrak{X}^r(M)$ defining the system (5.1).
The domain of attraction of $p^\star$ is the set

$$\mathcal{D}(\varphi_X, p^\star) = \{p \in M : \lim_{t \to +\infty} \varphi_X^t(p) = p^\star\}. \tag{5.3}$$

With some abuse of notation one could also write $\mathcal{D}(\varphi_X, p^\star)$ as $\omega^{-1}(\{p^\star\})$. The
domain of attraction is also called the *basin-* or *region* of attraction and is in other
work occasionally denoted as $B(p^\star)$ or $A(p^\star)$. Estimating the domain of attraction
has a variety of applications, for example in cancer treatments [51, 58]. See also
the 1985 survey paper by Genesio, Tartaglia and Vicino for more historical con-

---

[4] Given a continuous vector field $\dot{x} = f(x)$ with $f(0) = 0$ on $\mathbb{R}^n$, then, the origin is said to be
exponentially stable when there are $C, \lambda > 0$ such that $\|\varphi^t(x_0)\| \leq Ce^{-\lambda t}\|x_0\|$ for all $x_0 \in \mathbb{R}^n$.

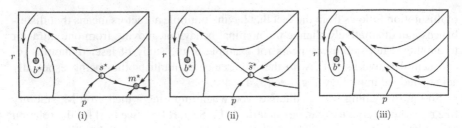

Fig. 5.2 Continuous transformation of Stepanova's model, removing the malignant equilibrium

text [23]. For general attractors $A \subseteq M$, one generalizes Definition 5.4 consistently, that is, $\mathcal{D}(\varphi_X, A) = \{p \in M : \lim_{n \to +\infty} \varphi_X^{t_n}(p) = a,\ a \in A$, for some monotonically increasing sequence $\{t_n\}_{n \in \mathbb{N}} \subset \mathbb{R}$, with $\lim_{n \to \infty} t_n = +\infty\}$.

In general, an equilibrium point $p^* \in M$ of some vector field $X \in \mathfrak{X}^r(M)$ is not locally asymptotically stable. In this case it might be of interest to split M, locally, in stable and unstable parts. Assume that p^* is a hyperbolic equilibrium point under the flow φ_X and define the so-called *stable* and *unstable* "*manifolds*" of p^* by $W^s(\varphi_X, p^*) = \{p \in M : \omega(\varphi_X, p) = \{p^*\}\}$ and $W^u(\varphi_X, p^*) = \{p \in M : \alpha(\varphi_X, p) = \{p^*\}\}$. As p^* is hyperbolic one can split $T_{p^*}M$ as $T_{p^*}M = T_{p^*}W^s(\varphi_X, p^*) \oplus T_{p^*}W^u(\varphi_X, p^*)$. In fact, Tp^*M splits according to the generalized eigenvectors of DX_{p^*}. More can be said about these stable- and unstable manifolds, see [33, Chap. 6]. Also, when p^* is not hyperbolic one can appeal to *center manifold theory*, e.g., see [13, 29].

This section ends with two explicit vector field examples from biology.

Example 5.1 (*Tumor immune interactions*) Returning to Chap. 3, in particular, Hopf's result (Proposition 3.6) tell us that indices adding up to 0 can be effectively "*homotoped away*". This observation is of use when one wants to know if a certain dynamical systems can be "*continuously deformed*" into another dynamical system. Here, we look at an example that models tumor immune system interactions [58, Chap. 8]. *Stepanova's model* is given by the following set of equations

$$\begin{cases} \dot{p} = \xi p F(p) - \theta p r \\ \dot{r} = \alpha(p - \beta p^2)r + \gamma - \delta r, \end{cases} \tag{5.4}$$

where p represents *tumor volume*, r *the immunocompetent cell density*, $F(p)$ a *growth rate* and all other parameters are constant coefficients. For common choices of parameters, (5.4) has two stable equilibrium points, a *benign* state b^* and a *malignant* state m^*, separated by a saddle s^*, see Fig. 5.2(i). A question of interest is if (5.4) can be deformed such that only the asymptotically stable benign equilibrium state prevails. Using the theory of vector field indices we see that s^* and m^* have indices of opposite sign such that they can be morphed into $\widetilde{s}^*$, see Fig. 5.2(ii). Similarly, as this intermediate equilibrium point has index 0, it can be removed completely, see Fig. 5.2(iii). Indeed, as b^* has index 1, as is the Euler characteristic of the rectangular

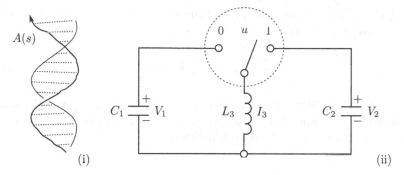

Fig. 5.3 Dynamical (control) systems: (i) a parametrization of a double helix DNA model on $SE(3, \mathbb{R})$ corresponding to Example 5.2 and (ii) the circuit corresponding to Example 5.4

domain, and the vector field is pointing inwards, the existence of this transformation was guaranteed from the start, see also Sect. 8.2.

Example 5.2 (*DNA conformation* [37]) In this example we model the DNA double helix as an elastic rod that has a helical twist at its minimum energy conformation. One can parametrize this model by a coordinate frame moving along a curve in $\mathbb{R}^3$. Moreover, this curve has a, possibly arc-length-dependent, matrix attached to it, describing its mechanical properties like stiffness. Now, to move along this curve, the instantaneous change in orientation and position need to be supplied, i.e., a rotation and a translation. This parametrization naturally leads to the employment of the *special Euclidean group*, defined as

$$SE(3, \mathbb{R}) = \left\{ A \in \mathbb{R}^{4 \times 4} : A = \begin{bmatrix} R & t \\ 0_{1 \times 3} & 1 \end{bmatrix}, R \in SO(3, \mathbb{R}), t \in \mathbb{R}^3 \right\}.$$

Here, for a $A \in SE(3, \mathbb{R})$, R describes the *rotation* and t the *translation*. For any point $p \in \mathbb{R}^3$, the propagation under $A \in SE(3, \mathbb{R})$ is as follows, append 1 to p, i.e., let $\tilde{p} = (p, 1) \in \mathbb{R}^4$, then $A\tilde{p} = (Rp + t, 1) \in \mathbb{R}^4$, i.e., first p is rotated to Rp, then, this point is translated to $Rp + t$. Now, the potential function that characterizes the energy of the DNA conformation will only consider *relative* changes in the curve parametrizing the DNA model. As such, we will always look from the so-called "*body frame*". To that end, see that for any differentiable curve $s \mapsto A(s) \in SE(3, \mathbb{R})$

$$A(s)^{-1} \frac{d}{ds} A(s) = \begin{bmatrix} R(s)^\mathsf{T} \dot{R}(s) & R(s)^\mathsf{T} \dot{t}(s) \\ 0_{1 \times 3} & 0 \end{bmatrix} \in \mathfrak{se}(3, \mathbb{R}) = T_{I_4} SE(3, \mathbb{R}).$$

With this observation in mind, let ξ be the vectorization of $A^{-1} \dot{A}$, denoted $\xi = (A^{-1} \dot{A})^\vee$ with inverse $\xi^\wedge = A^{-1} \dot{A}$, details can be found in [52], but are irrelevant for this example. Now, one defines a quadratic elastic potential energy function, for a DNA double helix modelled as an unconstrained extensible rod, by $U(\xi) =$

$\frac{1}{2}\xi^{\mathsf{T}} K \xi - k^{\mathsf{T}} \xi + \beta$, for some positive definite $K \in \mathbb{R}^{6 \times 6}$, $k \in \mathbb{R}^6$ and $\beta \in \mathbb{R}$. Minimizing U over ξ results in $\xi^\star(s) = K^{-1}k$ and therefore

$$\frac{d}{ds} A(s) = A(s)(K^{-1}k)^\wedge, \tag{5.5}$$

is a vector field on $\mathsf{SE}(3, \mathbb{R})$, that describes, locally, a DNA double helix model corresponding to the pair (K, k). To appreciate the geometric construction, the reader is invited to find the curve related to (5.5) directly.

5.2 Lyapunov Stability Theory

In this part we provide a brief overview of the stability theory as devised by Lyapunov [46]. The tools are fruitful in their own right, but as it turns out, the levelsets of Lyapunov functions are intimately related to the domain of attraction of the attractor at hand. A fairly general result is the following.

Theorem 5.1 (Compact attractors and Lyapunov functions [8, Theorem 10.6]) *Let φ be a continuous flow on a locally compact Hausdorff space X. The compact set $A \subseteq \mathsf{X}$ is a local attractor under φ if and only if there is continuous function $V : \mathcal{D}(\varphi, A) \to \mathbb{R}_{\geq 0}$ such that*

- (i) $V(x) = 0$ for all $x \in A$;
- (ii) $V(x) > 0$ for all $x \in \mathcal{D}(\varphi, A) \setminus A$, the sublevel set $V^{-1}([0, c]) = \{x \in \mathcal{D}$ $(\varphi, A) : V(x) \leq c\}$ is compact for all $c \in [0, +\infty)$ and;
- (iii) $V(\varphi^t(x)) < V(x)$ for all $x \in \mathcal{D}(\varphi, A) \setminus A$ and $t > 0$.

A function V as in Theorem 5.1 is called a *strict **continuous Lyapunov function***. The word "*strict*" refers to item (iii), i.e., the inequality is strict. Without strictness, attractivity of A cannot be asserted, merely Lyapunov stability. As we almost exclusively consider attractors, the adjective "*strict*" is dropped in the remainder of the book. See that the sublevel set compactness assures the V is (weakly) *coercive*, e.g., for $x \to \partial \mathcal{D}(\varphi, A)$ such that $\|x\| \to +\infty$, $V(x) \to +\infty$. See for example [26, p. 109] for more on the necessity of this condition, see also Example 5.3 for a geometric interpretation. Note that the results in [8] are with respect to (local) *semi*-dynamical systems, however, by Assumption 5.1 we do not need to concern ourselves with semidynamical system technicalities like "*start points*". See also [9, Chap. V] for a variety of Lyapunov theory, albeit for locally compact metric spaces. Although outside of the scope of this work, Theorem 5.1 does not apply as is to infinite-dimensional system, for infinite-dimensional Lyapunov theory, see for example [50].

Moreover, the results due to Kurzweil [38, Theorem 7], Massera [48] and Wilson [72] indicate that there should always be a *smooth* and proper[5] Lyapunov function. However, in contrast to popular belief, it took until the work by Fathi and Pageault

---

[5] That is, the sublevel sets of V are compact.

in 2019 to formally show this for flows generated by vector fields on smooth manifolds[6] [21]. See the proof of Theorem 6.2 for an application.

The key benefit of a *smooth* Lyapunov function is that explicit knowledge of the flow in Theorem 5.1(iii) can be substituted by a simpler condition. Let X be a continuous complete vector field on a smooth manifold M and let $A \subseteq M$ be an attractor under the flow φ_X. Then, there is a smooth and proper function $V : \mathcal{D}(\varphi_X, A) \to \mathbb{R}_{\geq 0}$ such that $A = \{x \in \mathcal{D}(\varphi_X, A) : V(x) = 0\}$ and the Lie derivative[7] $L_X V(x) < 0$ for all $x \in \mathcal{D}(\varphi_X, A) \setminus A$. Such a function V is called a ***smooth Lyapunov function***, again omitting "*strict*". For example, for $\dot{x} = -x^3$ one readily verifies that $V(x) = \frac{1}{2}x^2$ is a smooth Lyapunov function that asserts global asymptotic stability of $x^\star = 0$. See [15, 65] for further generalizations and see [35] for a survey on converse Lyapunov theorems. We also point out that the global *continuous* converse problem was solved by Conley and coworkers and is often referred to as "*Conley's Fundamental Theorem of Dynamical Systems*". The theorem states that "*Any flow on a compact metric space decomposes into a chain recurrent part and a gradient-like part.*", see [54]. We return to this in Sect. 8.5.

Example 5.3 (*Topology of a Lyapunov function* [73, Theorem 1.2]) Consider a continuous vector field $\dot{x} = f(x)$ on $\mathbb{R}^n$, with $f(0) = 0$ such that 0 is globally asymptotically stable and let $V : \mathbb{R}^n \to \mathbb{R}_{>0}$ be the corresponding smooth Lyapunov function. By assumption, $\langle \partial_x V(x), f(x) \rangle < 0$ on $\mathbb{R}^n \setminus \{0\}$, such that for any $c \in (0, +\infty)$ one crosses the level set $V_c = V^{-1}(c)$ once. Moreover, as we have the homeomorphism $\varphi_f^t(V_c) \simeq_t \varphi_f^s(V_c)$ for all $t, s \in \mathbb{R}$, we find by exploiting the flow properties of φ_f that $V_c \times \mathbb{R} \simeq_t \mathbb{R}^n \setminus \{0\}$. However, as the trivial bundle $V_c \times \mathbb{R}$ is homotopic to V_c and $\mathbb{R}^n \setminus \{0\} \simeq_h \mathbb{S}^{n-1}$ we can conclude that $V_c \simeq_h \mathbb{S}^{n-1}$. Better yet, as V_c is compact, the resolution of the (a) Poincaré conjecture yields that $V_c \simeq_t \mathbb{S}^{n-1}$. See also [38, Sect. 5] for earlier topological remarks and for example [23, Theorem 2] for early remarks on ramifications of $V_c \simeq_t \mathbb{S}^{n-1}$.

Finding explicit Lyapunov functions is usually *hard* [4] and contradicts being a (direct) "*method*" [32], but as upcoming sections illustrate, their mere existence provides to be useful. Moreover, as Theorem 5.1 holds for locally compact Hausdorff spaces, and not just topological manifolds, one can generalize a few upcoming Lyapunov-based results beyond topological manifolds indeed.

5.3 Control Systems

We start by illustrating how the study of a control system on a manifold can emerge.

Example 5.4 (*DC Converter* [75, Sect. 3.5], [44]) We will consider a so-called *DC-to-DC converter* as employed in laptops, phones and so forth. An idealized model can

---

[6] In particular, see [21, Sect. 1, Sect. 6] and [65] for clarifications.

[7] See [43, Chap. 9] for more on Lie derivatives.

be constructed as in Fig. 5.3, here the switch is used to either charge the left or right capacitor, by means of the inductor, thereby, one can control the potential (voltage) over these capacitors. The equations of motion follow from $V_L(t) = L(d/dt)I_L(t)$, $I_C(t) = C(d/dt)V_C(t)$ and Kirchoff's laws

$$\frac{d}{dt}\begin{bmatrix} C_1 V_1(t) \\ C_2 V_2(t) \\ L_3 I_3(t) \end{bmatrix} = \begin{bmatrix} (1-u)I_3(t) \\ u I_3(t) \\ -(1-u)V_1(t) - u V_2(t) \end{bmatrix},$$

where C_1, C_2 denote the *capacitance*, L_3 the *inductance* and $u \in \{0, 1\}$ the switching signal, that is, the input to the system. Now, consider the change of coordinates $x_1 = C_1^{1/2}V_1$, $x_2 = C_2^{1/2}V_2$ and $x_3 = L_3^{1/2}I_3$ with $x = (x_1, x_2, x_3) \in \mathbb{R}^3$. Then let $E(x) = \frac{1}{2}\langle x, x \rangle$ and observe that $d/dt E(x(t)) = 0$. This means that if the system starts from some initial condition $x_0 \in \mathbb{R}^3$, total energy is *preserved* and the system evolves on a 2-sphere with radius $(2E(x_0))^{1/2}$. Now, the reader is invited the draw qualitative conclusions from $\chi(\mathbb{S}^2) \neq 0$. We also remark that although the switching signal is binary, in practice one does not use such an idealized switch but rather a type of *transistor*. As such, u is not binary but rather *continuous*.

Example 5.5 (*Quantum control* [20]) Let $|\psi(t)\rangle \in \mathcal{H}$ be the state of a quantum system at time t, for some complex Hilbert space $\mathcal{H}$. Assume we work with a two-level system, that is, $|\psi(t)\rangle = c_0(t)|0\rangle + c_1(t)|1\rangle$, for $|0\rangle, |1\rangle \in \mathcal{H}$ and both $t \mapsto c_0(t)$ and $t \mapsto c_1(t)$ are complex-valued functions satisfying $|c_0(t)|^2 + |c_1(t)|^2 = 1$ for all t, i.e., the state is normalized. Let H_0 be the internal Hamiltonian and let the external part be of the form $\sum_{k=1}^m H_k \mu_k(t)$, for all H_i Hermitian operators on $\mathcal{H}$ and $t \mapsto \mu_k(t)$ real-valued (input) functions. Then, Schrödinger's equation, i.e., $i\hbar|\dot{\psi}(t)\rangle = H|\psi(t)\rangle$, becomes

$$i\hbar \frac{d}{dt}\Psi(t) = \left(\bar{H}_0 + \sum_{k=1}^m \bar{H}_k \mu_k(t)\right)\Psi(t), \tag{5.6}$$

for $\Psi(t)$ the *evolution operator* and $\bar{H}_i$ the Hamiltonian in (c_0, c_1) coordinates. One might be interested in steering $\Psi(0) = I_2$ to some desirable operator at some time $T \geq 0$. With this goal in mind, we can without loss of generality assume that all $\bar{H}_i$ are *traceless*. By doing so, we merely ignore a physically indistinguishable phase difference. Now see that (5.6) is of the form $\dot{\Psi}(t) = \bar{A}\Psi(t)$ with $\bar{A} = -i/\hbar \bar{H}$. However, now also consider the *special unitary group* $SU(n, \mathbb{C}) = \{Z \in \mathbb{C}^{n \times n} : Z^H Z = I_n\}$ with Lie algebra $\mathfrak{su}(n, \mathbb{C}) = \{A \in \mathbb{C}^{n \times n} : A^H + A = 0, \text{Tr}(A) = 0\}$ and see that $\bar{A} \in \mathfrak{su}(2, \mathbb{C})$. As $\Psi(0) = I_2$ is the identity element on $SU(2, \mathbb{C})$, we see that (5.6) is in fact a (right-invariant) control system on the Lie group $SU(2, \mathbb{C})$ cf. Sect. 1.3, unlocking Lie group machinery.

The purpose of the control system paradigm is to study if and how dynamical behaviour can be prescribed, e.g., the *stabilization problem* is that of finding inputs such that some set is stabilized in some sense. Control systems defined on manifolds can either be studied locally, that is, in some operating region of interest, or globally.

The most common *local* continuous-time formulation of a continuous time-invariant nonlinear dynamical control system is of the form

$$\Sigma_{n,m}^{\mathrm{loc}} : \left\{ \quad \frac{\mathrm{d}}{\mathrm{d}t} x(t) = f(x(t), u), \right. \tag{5.7}$$

for $x(t) \in \mathcal{X} \subseteq \mathbb{R}^n$ and $u \in \mathcal{U} \subseteq \mathbb{R}^m$ the state and input, respectively, e.g., see [53, 57]. One can append (5.7) with (time-invariant) output functions of the form $y(t) = h(x(t), u)$. These functions capture for example what one can measure. However, we will work directly with the state $x(t)$ and in that sense we restrict ourselves to a class of *open dynamical systems* with identity (or trivial) outputs [71]. This assumption is not restrictive as we will look at *obstructions* to achieving certain control objectives. Working with outputs $y(t)$ instead of all state variables $x(t)$ directly will only lead to more complications. We come back to this in Chap. 8.

When the input is chosen as a function of x, e.g., as $\mu(x) \in \mathcal{U}$, we speak of μ as being a *feedback* or *control law*. Note, as with the curve ξ and the points p and x, we use μ instead of u to differentiate between the function and the point. When μ depends on time, the input is said to be *time-varying*. A description like (5.7) works in Euclidean spaces or by using local coordinates on a manifold. However, as we are mostly interested in *global* questions we are aided by the following coordinate-free description, often attributed to Brockett, that allows for state-dependent input constraints, e.g., see the early work by Willems and van der Schaft [70, Definition 6]. Note, here we will not appeal to knowledge of a metric cf. [62, p. 123].

Definition 5.5 (*Continuous control system*) Given a smooth manifold M, a continuous control system is the triple $\Sigma = (\mathsf{M}, \mathcal{U}, F)$, consisting of a topological space $\mathcal{U}$, a continuous surjective map $\pi_u : \mathcal{U} \to \mathsf{M}$, the canonical projection $\pi_x : T\mathsf{M} \to \mathsf{M}$ and a continuous fiber-preserving map $F : \mathcal{U} \to T\mathsf{M}$ such that the following diagram is commutative:

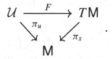

Definition 5.5 implies that the available inputs at $x \in \mathsf{M}$ are characterized by the sections $\Gamma^0(\mathcal{U})$, i.e., the continuous maps $\sigma : \mathsf{M} \to \mathcal{U}$ such that $\pi_u \circ \sigma = \mathrm{id}_{\mathsf{M}}$. Indeed, let $\mathcal{U} = \mathbb{R}^n \times \mathbb{R}^m$, then F relates directly to (5.7). More interestingly, consider $\mathcal{U}$ to be for example a *disk bundle* (tubular neighbourhood of M), which corresponds in coordinates to the input being constrained to a topological disk. In the framework of Definition 5.5, a continuous control law (feedback) $\mu \in \Gamma^0(\mathcal{U})$ results in the continuous *closed-loop* system $F \circ \mu : \mathsf{M} \to T\mathsf{M}$. Note that instead of assuming F and μ to be continuous maps, one could consider a larger class of systems by only demanding that the vector field $F \circ \mu$ is continuous. See also [24, 39, 53, 64].

Example 5.6 (*Case study Sect.* 1.3: *modelling*) Consider the left-invariant vector field (1.1b) over a lie group G. This model is analogous to $\dot{x} = u$ with $u \in \mathbb{R}^n$, as such, the control system becomes $\Sigma = (G, TG, \mathrm{id}_{TG})$.

Example 5.7 (*Affine control systems*) Motivated by mechanics, one of the most studied nonlinear dynamical control system models is of the *input-affine* form, cf. [53], described by a set of vector fields $\{f, g_1, \ldots, g_m\}$ and an input taking values in $\mathbb{R}^m$

$$\Sigma_{n,m}^{\mathrm{aff}} : \left\{ \quad \frac{\mathrm{d}}{\mathrm{d}t} x(t) = f(x(t)) + \sum_{i=1}^{m} g_i(x(t)) u_i. \right. \tag{5.8}$$

Globally, $\Sigma_{n,m}^{\mathrm{aff}}$ corresponds to $\pi_u : \mathcal{U} \to \mathsf{M}$ being a *vector* bundle [53, p. 428].

To put the class of continuous feedback laws in perspective, we must be able to describe *generic* admissible behaviour, that is, we must be able to describe *all* trajectories that can result from applying an admissible—and potentially discontinuous—input to the control system. Albeit its importance and attention received, this task, that falls under the umbrella of "*controllability*", is only partially solved. Loosely speaking, the control system $\Sigma = (\mathsf{M}, \mathcal{U}, F)$ is said to be **controllable** when for any $x_1, x_2 \in \mathsf{M}$ there is a finite $T \geq 0$ and an "*admissible input*" such that the "*resulting integral curve*" starting at x_1 arrives at x_2 in time T cf. [53, Definition 3.2], [57, Definition 11.1]. The notion of an admissible input depends evidently on the application at hand, but let us highlight common mathematical assumptions. To that end, consider momentarily a control system in local coordinates, when studying dynamical control systems of the form $\dot{x} = f(x, u)$, the input might depend on time in a merely *measurable* (i.e., possibly discontinuous) way. As such, an integral curve that satisfies this differential equation cannot always be *continuously* differentiable.

To analyze this scenario, let $f : \mathbb{R} \times \Omega \to \Omega$ be continuous with Ω open in $\mathcal{X}$ and rewrite the standard initial value problem, that is, finding a curve ξ such that

$$\frac{\mathrm{d}}{\mathrm{d}t} \xi(t) = f(t, \xi(t)), \quad \xi(t_0) = \xi_{t_0}, \quad t_0 \in \mathrm{int}(\mathcal{I}), \quad \text{for } t \in \mathcal{I} \subseteq \mathbb{R}$$

as that of finding a curve $\xi : \mathcal{I} \to \Omega$ such that $\xi(t) = \xi_{t_0} + \int_{t_0}^{t} f(\tau, \xi(\tau)) \mathrm{d}\tau$ for all $t \in \mathcal{I}$. The integral representation of the initial value problem immediately reveals that demanding ξ to be C^1-smooth is not a necessity. Moreover, from a practical point of view, the integral problem is as relevant as the differential equation. Therefore, the right[8] setting is that of **absolutely continuous** curves, that is, curves $\xi : \mathcal{I} \to \Omega$ that, when restricted to compact subsets of $\mathcal{I}$, admit an integral representation of the form $\xi(t) = \xi(t_0) + \int_{t_0}^{t} g(\tau) \mathrm{d}\tau$ for some Lebesgue measurable function $g : \mathcal{I} \to \Omega$ with $t \in [t_0, t_1] \subseteq \mathcal{I}$. Then, one speaks of a (weak) solution to the initial value problem when one finds an absolutely continuous curve $\xi : \mathcal{I} \to \Omega$, that passes through $\xi_{t_0} \in \Omega$ at $t_0 \in \mathcal{I}$, such that $\dot{\xi}(t) = f(t, \xi(t))$ holds for *almost every* $t \in \mathcal{I}$ in the sense of

---

[8] Although even weaker regularity notions can be desirable, see Sect. 7.3.

Lebesgue. To assert existence one commonly appeals to **Carathéodory's sufficient conditions**, that is, $t \mapsto f(t, \xi)$ must be measurable for all ξ and $\xi \mapsto f(t, \xi)$ must be continuous for (almost) all t. This implies in particular that $t \mapsto f(t, \xi(t))$ is measurable. Moreover, for each compact set $K \subset \mathcal{I} \times \Omega$ there must be an integrable function $b_K(t)$ such that $\| f(t, \xi) \| \le b_K(t)$ for all $(t, \xi) \in K$ [27, Sect. 1.5].

Now we return to dynamical control systems and briefly comment on classes of inputs that are frequently employed. The input $t \mapsto \mu(t) \in \mathcal{U}$ with $t \in \mathcal{I}$ is said to be **locally integrable** when $\int_{\mathcal{J}} \| \mu(\tau) \| d\tau < \infty$ for all compact subsets $\mathcal{J} \subseteq \mathcal{I}$ (with $\| \cdot \|$ from $\mathbb{R}^m$). We denote this by $\mu \in L^1_{\text{loc}}(\mathcal{I})$. On the other hand $t \mapsto \mu(t) \in \mathcal{U}$ is said to be **locally essentially bounded** when for each compact set $\mathcal{J} \subseteq \mathcal{I}$ there is a compact set $K \subseteq \mathcal{U}$ such that $\mu(t) \in K$ for almost every $t \in \mathcal{J}$. This set is denoted by $L^\infty_{\text{loc}}(\mathcal{I})$. When the control system is input-affine, we can select $\mu \in L^1_{\text{loc}}(\mathcal{I})$ to assert, locally, the Carathéodory conditions, notably, the last condition. In general, when $f(x, u)$ is jointly continuous in x and u one select $\mu \in L^\infty_{\text{loc}}(\mathcal{I})$ [62, Appendix C].

With this motivation and the control system $\Sigma = (M, \mathcal{U}, F)$ in mind, we define the set of **admissible inputs** over the interval $\mathcal{I}$, denoted by $\mathcal{U}(\mathcal{I})$, as all maps $\mu : (\mathcal{I} \subseteq \mathbb{R}) \times M \to \mathcal{U}$ with $t \mapsto \mu(t, x)$ being measurable for all $x \in M$, $x \mapsto \mu(t, x)$ being continuous for all $t \in \mathcal{I}$ and in particular $\mu(t, \cdot) \in \Gamma^0(\mathcal{U})$ such that **controlled trajectories** are absolutely continuous curves $\xi : \mathcal{I} \to M$ such that

$$\frac{d}{dt} \xi(t) = (F \circ \mu)(t, \xi(t)) \quad \text{for u.e. } t \in \mathcal{I} \text{ and some } \mu \in \mathcal{U}(\mathcal{I}).$$

With this in place we return to the question of controllability. Global controllability is unfortunately difficult to characterize. To study controllability and its ramifications, locally, we follow Lewis [45] and define the reachable set from $x_1 \in M$ in time $T \ge 0$ as $\mathcal{R}_\Sigma(x_1, T) = \{ \xi(T) :$ there exists a controlled trajectory $\xi : \mathcal{I} \to M$ such that $\xi(0) = x_1$ and $[0, T] \subseteq \mathcal{I}. \}$. Let $\mathcal{R}_\Sigma(x, T) = \cup_{t \in T} \mathcal{R}_\Sigma(x, t)$, then we say that Σ is **accessible** from $x \in M$ if $\text{int}(\mathcal{R}_\Sigma(x, \mathbb{R}_{\ge 0})) \ne \emptyset$. Similarly, Σ is **locally controllable** from $x \in M$ when $x \in \text{int}(\mathcal{R}_\Sigma(x, \mathbb{R}_{\ge 0}))$. The control system Σ is **small-time-locally-controllable** from $x \in M$ if there is a $T > 0$ such that $x \in \text{int}(\mathcal{R}_\Sigma(x, [0, s]))$ for all $s \in (0, T]$. Based on the control system at hand we will spell out—as far as possible—which type of controllability can be asserted and how. When the goal is to stabilize a set $A \subseteq M$ one might make a distinction between controllability on $M \setminus A$ and on A itself. See [45] for a relatively recent overview of geometric nonlinear controllability and see [34] for earlier survey paper by Kawski. See also [22] for the parallel development of exploiting so-called "*flatness*".

Example 5.8 (*Linear controllability and continuous feedback*) Consider the linear control system $\Sigma^L = (\mathbb{R}^n, \mathbb{R}^n \times \mathbb{R}^m, F \in L(\mathbb{R}^n \times \mathbb{R}^m; \mathbb{R}^n))$, succinctly given by

$$\dot{x} = Ax + Bu, \tag{5.9}$$

for some $A \in \mathbb{R}^{n \times n}$ and $B \in \mathbb{R}^{n \times m}$. Note, here one exploits the identification $T\mathbb{R}^n \simeq \mathbb{R}^n \times \mathbb{R}^n$. It readily follows that the integral curves $t \mapsto \xi(t) \in \mathbb{R}^n$ corresponding to (5.9) are of the form $\xi(T) = e^{TA}\xi(0) + \int_0^T e^{(T-s)A} B\mu(s)ds$ for some $T > 0$ and some choice of admissible input $t \mapsto \mu(t)$, e.g., $\mu \in L^1_{\text{loc}}([0, T])$. The linear control system Σ^L is controllable if and only if rank$(B, AB, \ldots, A^{n-1}B) = n$, which readily follows from comparing the rank of $(B, AB, \ldots, A^{n-1}B)$ to the rank of $e^{tA}B$ cf. [62, Theorem 3]. Instead of referring to Σ^L, one frequently refers to the pair (A, B), which is said to be a controllable pair in case Σ^L is controllable. Now, one can show that if (A, B) is a controllable pair, so is $(A + BK, B)$, for any $K \in \mathbb{R}^{m \times n}$ [74, Lemma 2.1]. Better yet, one can show that there is a $K_0 \in \mathbb{R}^{m \times n}$ and a $u_0 \in \mathbb{R}^m$ such that $(A + BK_0, Bu_0)$ is a controllable pair [74, Lemma 2.2]. Note, with some abuse of notation, the resulting control system $\dot{x} = (A + BK_0)x + (Bu_0)u$ is now a controllable *single-input* system. Then, for the moment assume $m = 1$, $B = b$ and that (A, b) is a controllable pair, i.e., the matrix $T = (b, Ab, \ldots, A^{n-1}b)$ has rank n. Due to the Cayley–Hamilton theorem, performing a linear change of coordinates $Tz = x$ can be shown to result in

$$T^{-1}AT = \widetilde{A} = \begin{bmatrix} 0 & 0 & \cdots & -a_1 \\ 1 & 0 & \ldots & -a_2 \\ 0 & \ddots & & \vdots \\ 0 & 0 & 1 & -a_n \end{bmatrix}, \quad T^{-1}b = \widetilde{b} = \begin{bmatrix} 1 \\ 0 \\ \vdots \\ 0 \end{bmatrix}. \tag{5.10}$$

for some tuple $(a_1, \ldots, a_n)$. Then, consider the pair $(\bar{A}, \bar{b})$ given by

$$\bar{A} = \begin{bmatrix} -\bar{a}_1 & -\bar{a}_2 & \cdots & -\bar{a}_n \\ 1 & 0 & \ldots & 0 \\ 0 & \ddots & & \vdots \\ 0 & 0 & 1 & 0 \end{bmatrix}, \quad \bar{b} = \begin{bmatrix} 1 \\ 0 \\ \vdots \\ 0 \end{bmatrix}.$$

A pair of the form $(\bar{A}, \bar{b})$ is controllable for any tuple $(\bar{a}_1, \bar{a}_2, \ldots, \bar{a}_n)$ and is said to be in the *control canonical form*. Hence, as any controllable pair can be transformed as in (5.10), there must exist an invertible linear transformation S such that any controllable pair (A, b) can be brought into this control canonical form [67, Chap. 3]. To that end, assume that $(\bar{A}, \bar{b})$ is a controllable pair in canonical form and let $\bar{k} \in \mathbb{R}^{1 \times n}$, then, by employing the feedback $\mu(x) = \bar{k}x$ one finds that the characteristic polynomial of $\bar{A} + \bar{b}\bar{k}$ is of the form $\lambda^n + (\bar{k}_1 - a_1)\lambda^{n-1} + \cdots + (\bar{k}_n - a_n)$. Therefore, one can "*place*" the eigenvalues of $\bar{A} + \bar{b}\bar{k}$ anywhere desired by an appropriate selection of $\bar{k}$. Concluding, we have shown that for any controllable linear dynamical system Σ^L there is a tuple $(K_0, u_0, \bar{k}, S^{-1})$ such that the linear feedback $\mu(x) = Kx = (K_0 + u_0\bar{k}S^{-1})x$ globally asymptotically stabilizes $x^\star = 0$.

In general, however, controllability as is does not provide any information on *how* the input should be selected, e.g., the input space might contain any measurable function. To that end, we highlight a variety of *continuous* stabilization paradigms with respect to a continuous control system $\Sigma = (M, \mathcal{U}, F)$.

(i) *Static feedback*: Control laws are continuous sections $\sigma : \mathsf{M} \to \mathcal{U}$.

(ii) *Dynamic feedback*: This is most easily described in coordinates by passing from static feedback in u, i.e., $\dot{x} = f(x, u)$ to dynamic feedback in u and v, i.e., $\dot{x} = f(x, u), \dot{z} = v$. Here, both inputs can depend on x and the auxiliary state-variable z. In the context of input-output systems one commonly encounters the more restrictive form $\dot{x} = f(x, z), \dot{z} = v$.

(iii) *Time-varying controls*: Similar to dynamic feedback, one can describe a time-varying vector field by lifting the state space, that is, by writing $\dot{x} = f(t, x)$ as $\dot{x} = f(s, x), \dot{s} = 1$. In this case, the time variable is understood to live in a subset of $\mathbb{R}$. If instead the time variable lives on $\mathbb{S}^1$ the system is *periodic*, e.g., $\dot{s} = -Js$ for $J \in \mathsf{Sp}(2, \mathbb{R})$ (the *Symplectic group*) corresponding to a clockwise rotation. Crucially, the introduction of the auxiliary variable s will require a different analysis as $\dot{s} \neq 0$ and solutions generally diverge to $+\infty$, see Example 6.4.

As highlighted in the introduction, in what follows we focus on continuous feedback laws due to implementation and robustness considerations, but also to investigate the limitations of this common assumption. As dynamic and time-varying controls can be modelled by means of vector fields on extended state spaces, this is a good initial vantage point. However, keeping these distinctive classes in mind is important. For example, Coron and Praly showed that there are systems that cannot be stabilized by static feedback while a stabilizing dynamic controller exists [19]. This result relates to what is a common observation in topology, extending state spaces can enable continuity.[9] To give a simplified example on $\mathbb{R}^2$, for $f(x_1, x_2) = (x_1, -x_2)$ one has $\mathrm{ind}_0(f) = -1 \neq (-1)^n$, while for $\bar{f}(x_1, x_2, x_3) = (x_1, -x_2, x_3)$ one has $\mathrm{ind}_0(\bar{f}) = -1 = (-1)^{n+1}$ cf. Example 3.4. Moreover, Coron showed that stabilizing time-varying controllers are significantly more rich in that they usually exist [16, 17]. We have more to say about this in the upcoming two sections. See also [68] for more on the application of dynamic feedback. Note, in all of the aforementioned we focus on *state*-feedback, meaningful extensions to non-trivial outputs are still open. In the past the focus was mostly on local stabilization of equilibrium points, i.e., a controller induces a well-behaved closed-loop system on a neighbourhood of the equilibrium point. We like to understand what could happen outside of this operating region. Exactly then, the *global* topology of the underlying space plays a critical role.

Example 5.9 (*Control-Lyapunov functions*) Assume to work with a input-affine control system of the form (5.8) and let $x^\star = 0$ be the point to be stabilized under a feedback μ that satisfies $\mu(x^\star) = 0$. If we can find a smooth and proper function $V : \mathbb{R}^n \to \mathbb{R}_{\geq 0}$ that only vanishes at 0 and is such that for all $x \in \mathbb{R}^n \setminus \{0\}$ there is a $u \in \mathbb{R}^m$ such that $\langle \partial_x V(x), f(x) + \sum_i g_i(x)u_i \rangle < 0$, then V is a *control-Lyapunov function* (CLF) [62, Lemma 5.7.4]. Such a construction is particularly interesting when the inputs are known to be constrained. Although controllability does not imply the existence of a stabilizing continuous feedback, when V is a CLF and additionally

---

[9] Consider embedding the figure 8 in $\mathbb{R}^2$ and in $\mathbb{R}^3$.

satisfies the *small control property*,[10] then, the work by Artstein [7] and Sontag [60, 61] shows that an explicit *continuous* stabilizing feedback law *can* be found. As will be shown, the existence of CLFs is thereby easily obstructed based on topological grounds. See also Sect. 7.3.

Remark 5.1 (*From stability to continuous stabilization*) As continuity is preserved under compositions, a common approach to provide topological obstructions to continuous stabilization is as follows. Assume one can show that there does not exist any continuous vector field $X \in \mathfrak{X}^{r \geq 0}(M)$ that satisfies property P. In its turn, this result directly implies that for any control system $\Sigma = (M, \mathcal{U}, F)$ in the sense of Definition 5.5, there is no continuous feedback $\mu : M \to \mathcal{U}$ such that the closed-loop vector field $F \circ \mu$ satisfies property P, e.g., there is no continuous feedback law on M that can achieve this type of stabilization.

For more on dynamical systems, see [8, 9, 31, 33, 55, 56, 59], for more on Lyapunov theory, see [9, 36, 62] and for more on geometric control theory, see [3, 11, 12, 18, 29, 53, 57]. See [3, Sect. 8] for a discussion on the topology of *attainable* sets, also called *reachable-* or *accessible* sets. Although we focus on smoothness and continuity, in part due to converse Lyapunov theorems, see [63] for an overview of how real-analyticity presents itself as an important setting for the study of control theory. We also remark that the work by Lur'e and Postnikov [47] was one of the first that—amongst other things—linked Lyapunov stability theory to control theory, whereas the book (and invariance principles) by Lefschetz and LaSalle [40] was key in promotion of the concept.

References

1. Abraham R, Marsden J, Ratiu T (1988) Manifolds, tensor analysis, and applications, 2nd edn. Applied Mathematical Sciences, Springer, New York
2. Abraham R, Marsden JE (2008) Foundations of mechanics. American Mathematical Society, Providence
3. Agrachev AA, Sachkov YL (2004) Control theory from the geometric viewpoint. Springer Science and Business Media, Berlin
4. Ahmadi AA, Krstic M, Parrilo PA (2011) A globally asymptotically stable polynomial vector field with no polynomial Lyapunov function. In: Proceedings of IEEE conference on decision and control, and European control conference, pp 7579–7580
5. Angeli D, Ferrell JE, Sontag ED (2004) Detection of multistability, bifurcations, and hysteresis in a large class of biological positive-feedback systems. Proc Natl Acad Sci USA 101(7):1822–1827
6. Arecchi F, Meucci R, Puccioni G, Tredicce J (1982) Experimental evidence of subharmonic bifurcations, multistability, and turbulence in a q-switched gas laser. Phys Rev Lett 49(17):1217
7. Artstein Z (1983) Stabilization with relaxed controls. Nonlinear Anal-Theor 7(11):1163–1173
8. Bhatia NP, Hájek O (2006) Local semi-dynamical systems. Springer, Berlin
9. Bhatia NP, Szegö GP (1970) Stability theory of dynamical systems. Springer, Berlin

---

[10] That is, for any $\varepsilon > 0$ there must be a $\delta > 0$ such that if $x \neq 0$ satisfies $\|x\| < \delta$ then there is a u with $\|u\| < \varepsilon$ satisfying the control-Lyapunov condition $L_f V(x) + \sum_i u_i L_{g_i} V(x) < 0$.

10. Birkhoff GD (1927) Dynamical systems. American Mathematical Society, Providence
11. Bloch A (2015) Nonholonomic mechanics and control. Springer, New York
12. Bullo F, Lewis AD (2004) Geometric control of mechanical systems. Springer, New York
13. Carr J (2012) Applications of centre manifold theory. Springer Science and Business Media, New York
14. Cheng C-Y, Lin K-H, Shih C-W (2006) Multistability in recurrent neural networks. SIAM J Appl Math 66(4):1301–1320
15. Clarke FH, Ledyaev YS, Stern RJ (1998) Asymptotic stability and smooth Lyapunov functions. J Differ Equ 149(1):69–114
16. Coron J-M (1992) Global asymptotic stabilization for controllable systems without drift. Math Control Sig Syst 5(3):295–312
17. Coron J-M (1995) On the stabilization in finite time of locally controllable systems by means of continuous time-varying feedback law. SIAM J Contr Optim 33(3):804–833
18. Coron J-M (2007) Control and nonlinearity. American Mathematical Society, Providence
19. Coron J-M, Praly L (1991) Adding an integrator for the stabilization problem. Syst Control Lett 17(2):89–104
20. D'alessandro D, Dahleh M (2001) Optimal control of two-level quantum systems. IEEE Trans Automat Contr 46(6):866–876
21. Fathi A, Pageault P (2019) Smoothing Lyapunov functions. T Am Math Soc 371(3):1677–1700
22. Fliess M, Lévine J, Martin P, Rouchon P (1995) Flatness and defect of non-linear systems: introductory theory and examples. Int J Control 61(6):1327–1361
23. Genesio R, Tartaglia M, Vicino A (1985) On the estimation of asymptotic stability regions: state of the art and new proposals. IEEE T Automat Contr 30(8):747–755
24. Grizzle J, Marcus S (1985) The structure of nonlinear control systems possessing symmetries. IEEE T Automat Contr 30(3):248–258
25. Grüne L, Sontag ED, Wirth FR (1999) Asymptotic stability equals exponential stability, and ISS equals finite energy gain-if you twist your eyes. Syst Control Lett 38(2):127–134
26. Hahn W (1967) Stability of motion. Springer, Berlin
27. Hale JK (1980) Ordinary differential equations. Krieger Publishing Company, Malabar
28. Hurley M (1982) Attractors: persistence, and density of their basins. T Am Math Soc 269(1):247–271
29. Isidori A (1985) Nonlinear control systems: an introduction. Springer, Berlin
30. Jafarpour S, Lewis AD (2014) Time-varying vector fields and their flows. Springer, Cham
31. Jost J (2005) Dynamical systems: examples of complex behaviour. Springer Science and Business Media, Berlin
32. Kalman RE, Bertram JE (1960) Control system analysis and design via the "second method" of Lyapunov: I-continuous-time systems. J Basic Eng-T ASME 82(2):371–393
33. Katok A, Hasselblatt B (1995) Introduction to the modern theory of dynamical systems. Cambridge University Press, New York
34. Kawski M (1990) High-order small-time local controllability. Nonlinear controllability and optimal control, vol 133. Dekker, New York, pp 431–467
35. Kellett CM (2015) Classical converse theorems in Lyapunov's second method. Discrete Cont Dyn-B 20(8):2333–2360
36. Khalil HK (2002) Nonlinear systems. Prentice Hall
37. Kim JS, Chirikjian GS (2006) Conformational analysis of stiff chiral polymers with end-constraints. Mol Simul 32(14):1139–1154
38. Kurzweil J (1963) On the inversion of Ljapunov's second theorem on stability of motion. AMS Transl Ser 2(24):19–77
39. Kvalheim MD, Koditschek DE (2022) Necessary conditions for feedback stabilization and safety. J Geom Mech
40. La Salle J, Lefschetz S (1961) Stability by Liapunov's direct method with applications. Academic Press, New York
41. Laurent M, Kellershohn N (1999) Multistability: a major means of differentiation and evolution in biological systems. Trends Biochem Sci 24(11):418–422

42. Lee JM (1997) Riemannian manifolds. Springer, New York
43. Lee JM (2012) Introduction to smooth manifolds. Springer, New York
44. Leonard NE, Krishnaprasad P (1994) Control of switched electrical networks using averaging on Lie groups. In: Proceedings of IEEE conference on decision and control, vol 2, pp 1919–1924
45. Lewis AD (2001) A brief on controllability of nonlinear systems
46. Liapunov A (1892) A general task about the stability of motion. Dissertation, University of Kharkov
47. Lur'e AI, Postnikov VN (1944) On the theory of stability of control systems. Appl Math Mech 8(3):246–248
48. Massera JL (1956) Contributions to stability theory. Ann Math 64(1):182–206
49. May RM (1977) Thresholds and breakpoints in ecosystems with a multiplicity of stable states. Nature 269(5628):471–477
50. Mironchenko A, Wirth F (2019) Non-coercive Lyapunov functions for infinite-dimensional systems. J Differ Equ 266(11):7038–7072
51. Moussa K, Fiacchini M, Alamir M (2021) Robust domain of attraction estimation for a tumor growth model. Appl Math Comput 410:126482
52. Murray RM, Li Z, Sastry SS (1994) A mathematical introduction to robotic manipulation. CRC Press, Boca Raton
53. Nijmeijer H, van der Schaft A (1990) Nonlinear dynamical control systems. Springer, New York
54. Norton DE (1995) The fundamental theorem of dynamical systems. Comment. Math. U. Carolinae 36(3):585–597
55. Palis JJ, De Melo W (1982) Geometric theory of dynamical systems: an introduction. Springer, New York
56. Robinson C (1995) Dynamical systems: stability, symbolic dynamics, and chaos. CRC Press, Boca Raton
57. Sastry S (1999) Nonlinear systems. Springer, New York
58. Schättler H, Ledzewicz U (2015) Optimal control for mathematical models of cancer therapies. Springer, New York
59. Shub M (1987) Global stability of dynamical systems. Springer Science and Business Media, Berlin
60. Sontag ED (1983) A Lyapunov-like characterization of asymptotic controllability. SIAM J Control Optim 21(3):462–471
61. Sontag ED (1989) A 'universal' construction of Artstein's theorem on nonlinear stabilization. Syst Control Lett 13(2):117–123
62. Sontag ED (1998) Mathematical control theory: deterministic finite dimensional systems. Springer, New York
63. Sussmann H (1990) Why real analyticity is important in control theory. Perspectives in control theory. Birkhäuser, Boston, pp 315–340
64. Tabuada P, Pappas GJ (2005) Quotients of fully nonlinear control systems. SIAM J Contr Optim 43(5):1844–1866
65. Teel AR, Praly L (2000) A smooth Lyapunov function from a class-$\mathcal{KL}$ estimate involving two positive semidefinite functions. ESAIM Contr Optim Ca 5:313–367
66. Teel AR, Zaccarian L (2006) On 'uniformity' in definitions of global asymptotic stability for time-varying nonlinear systems. Automatica 42(12):2219–2222
67. Trentelman HL, Stoorvogel AA, Hautus M (2001) Control theory for linear systems. Springer-Verlag, London
68. Tsuzuki T, Yamashita Y (2008) Global asymptotic stabilization for a nonlinear system on a manifold via a dynamic compensator. IFAC Proc Vol 41(2):6178–6183 (17th IFAC World Congress)
69. Vidyasagar M (2002) Nonlinear systems analysis. Society for Industrial and Applied Mathematics, Philadelphia
70. Willems J, Van der Schaft A (1982) Modelling of dynamical systems using external and internal variables with applications to Hamilton systems. Dynamical Systems and Microphysics. Academic Press, New York, pp 233–264

71. Willems JC (1998) Open dynamical systems and their control. Doc Math Extra ICM(III):697–706
72. Wilson FW (1969) Smoothing derivatives of functions and applications. T Am Math Soc 139:413–428
73. Wilson FW Jr (1967) The structure of the level surfaces of a Lyapunov function. J Differ Equ 3(3):323–329
74. Wonham WM (1979) Linear multivariable control. Springer, New York
75. Wood JR (1974) Power conversion in electrical networks. PhD thesis, Harvard University

Chapter 6
Topological Obstructions

Given a continuous control system $\Sigma = (\mathsf{M}, \mathcal{U}, F)$, or merely the manifold M, in this chapter we consider stability and stabilization on three levels.

(i) What is the topology of an admissible attractor $A \subseteq \mathsf{M}$?
(ii) Given an attractor A, what is the topology of the domain of attraction $\mathcal{D} \subseteq \mathsf{M}$?
(iii) Given the pair $(A, \mathcal{D})$, what kind of dynamics are imposed on $\mathsf{M} \setminus \mathcal{D}$?

Regarding the notions of stability and stabilization, we primarily focus on (uniform) asymptotic stability and continuous static feedback, but note in passing if these obstructions prevail under other stabilization paradigms. To summarize the list of above, see Fig. 6.1, we want to understand how the topology of M influences stability and stabilization possibilities.

We will address Questions (i)–(iii) above, by looking at the stabilization of *points*, *submanifolds* and generic *sets*. The obstructions differ in being of a *local* or *global* nature.[1] Moreover, some require knowledge of the vector field (dynamical system) at hand, whereas other obstruction hold for example for any continuous flow on M. For instance, with respect to the stabilization of some equilibrium point, an elementary necessary condition on Σ is that $F(\mathcal{U}) \cap Z_{\pi_p}(\mathsf{M}) \neq \emptyset$, or stronger, for a particular point $p^\star \in \mathsf{M}$ it must be true that $(p^\star, 0) \in F(\pi_u^{-1}(p^\star))$. Similarly, for the stabilization of some invariant submanifold $A \hookrightarrow \mathsf{M}$ one needs $F(\pi_u^{-1}(a)) \cap (a, T_a \mathsf{M}|_A) \neq \emptyset$ for all $a \in A$. For a generic compact attractor $A \subseteq \mathsf{M}$, however, one needs that for all $p \in \partial A$ the set $F(\pi_u^{-1}(p))$ does not only contain pairs (p, v) with v pointing outward of A, which is asserted in local coordinates. Similar conditions can be stated in the language of involutive distributions[2] on M [94]. It turns out that incorporating topological aspects of the control system Σ results in significantly deeper insights than the mere existence of equilibrium points and so forth.

---

[1] Here, the difference between these settings is best understood as that in the global setting one *does* have *explicit* knowledge of the domain of attraction, in contrast to the local setting.

[2] A fruitful concept, yet outside the scope of this work.

© The Author(s) 2023
W. Jongeneel and E. Moulay, *Topological Obstructions to Stability and Stabilization*,
SpringerBriefs in Control, Automation and Robotics,
https://doi.org/10.1007/978-3-031-30133-9_6

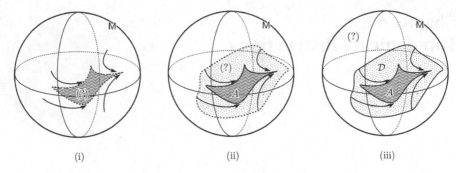

(i)   (ii)   (iii)

Fig. 6.1 Given M, (i) what is *the topology* (to be understood as the homotopy type) of the attractor; (ii) what is the topology of the domain of attraction; (iii) how does this fit into M, what kind of qualitative dynamics are imposed outside of $\mathcal{D}$?

One of the first principled overviews towards answering Question (ii) appeared in the book by Bhatia and Szegö [10, Sect. V.3]. Early comments on how the global topology in combination with the Poincaré–Hopf theorem imposes an obstruction for stabilization appeared in 1988 by Koditschek [71]. Similar comments appeared in [72] with respect to navigation problems and obstacle avoidance. As the focus is on generic results, we omit early specialized (low-dimensional) results, e.g., early work by Dayawansa,[3] see [38] and references therein. We also omit explicit discussions about obstructions due to quantization and the like.

With respect to Question (i), we mostly focus on compact attractors, the main reason being that asymptotic stability of a compact attractor can be characterized solely by topological means, cf. Sect. 5.1. Besides, a set of equilibrium points is closed [10, Theorem II 2.7], so that if M is compact, any subset $A \subseteq M$ on which some vector field vanishes is necessarily compact.[4] See also [79] for how compactness of A turned out to be a necessity in correcting a topological result due to Wilson.

6.1   Obstructions to the Stabilization of Points

Equipped with the tools from Chaps. 3–5, we start by considering the stabilization of an (equilibrium) point $p \in M$, either locally, or globally. We emphasize that locally asymptotically stable equilibrium points are isolated, yet, to aid the reader we do occasionally keep the adjective *isolated*.

---

[3] However, we like to emphasize that Dayawansa inspired a lot of work we do discuss.

[4] The proof of [10, Theorem II 2.7] relies on M being a metric space. However, under our assumptions, the result can be shown via embedding M in some Euclidean space $\mathbb{R}^d$ as well.

6.1.1 Local Obstructions

In this section we mostly focus on local nonlinear models of the form $\Sigma_{n,m}^{loc}$ (5.7) with $f(0,0) = 0$ and seek control strategies such that $x^\star = 0$ is an isolated and *locally* asymptotically stabilized equilibrium point.

We start with a few local necessary conditions for local asymptotic stabilization by means of continuous control laws. In particular, we look at Brockett's (degree) condition, Zabczyk's (index) condition and Coron's (homological) condition.

Consider some non-surjective map $g : X \to Y$, as there must be a point $y \in Y$ such that $g^{-1}(y) = \emptyset$, y is a regular value and $\deg(g) = 0$ cf. Sect. 3.5. Now, by Example 3.4 we know that isolated asymptotically stable equilibrium points have non-zero index $(-1)^n$. Then, it follows from Lemma 3.3 and Example 3.2 that (in the Euclidean case) $\text{ind}_0(X) \neq 0$ implies that the vector field X is locally surjective around 0. The reason being that one can homotope $X(p)/\|X(p)\|_2$ to $(X(p) - p')/\|X(p) - p'\|_2$ for sufficiently small p', preserving the degree and hence $X(p) = p'$ must have a solution for any sufficiently small p'. See also [131, Lemma 3]. Therefore, considering the control system $\Sigma_{n,m}^{loc}$, the map $f : \mathbb{R}^n \times \mathbb{R}^m \to \mathbb{R}^n$ must at least be locally surjective from a neighbourhood of $(0,0)$ onto a neighbourhood of 0 to allow for a non-zero vector field index. Note that this is a necessary but not sufficient condition for the degree to be non-zero.

Given these observations, Brockett's celebrated necessary condition follows.[5]

Theorem 6.1 (Brockett's condition [18, Theorem 1.(iii)]) *Let $\Sigma_{n,m}^{loc}$ be a local continuous control system. Then, there is a continuous feedback $x \mapsto \mu(x) \in \mathbb{R}^m$ with $\mu(0) = 0$ rendering $0 \in \mathbb{R}^n$ locally asymptotically stable only if $(x, u) \mapsto f(x, u)$ is a surjective map from a neighbourhood of $(0,0)$ onto a neighbourhood of 0.*

Proof (*Sketch*) When the control system is smooth, assume to have knowledge of a smooth controller μ with $\mu(0) = 0$ such that $f(x, \mu(x))$ is asymptotically stable. For $\text{ind}_0(f) \neq 0$ we clearly need $x \mapsto f(x, \mu(x))$ to be surjective.

See the proof of Theorem 6.2 for more on the purely continuous case, or see, [18, 116, Theorem 22]. In particular, after Zabczyk [131], Orsi et al. generalized Theorem 6.1 to the setting where the continuous vector field does not necessarily give rise to a (local) flow [95]. See also [12, Theorem 4] for a generalization to nonholonomic control systems. A composition operator theoretic study of feedback stabilization is undertaken by Christopherson, Mordukhovich and Jafari in [31]. Importantly, by building upon Hautus, Brockett and Zabczyk, their work goes beyond necessary conditions and towards *sufficient* conditions for asymptotic stabilization. See also that a bijective reparametrization of the input does not change the theorem qualitatively. As indicated in [18], the surprising element of Theorem 6.1 is that controllability does not seem to play a key role when it comes to sufficiency. Indeed, one can find examples of systems that *are* controllable, yet, not asymptotically stabilizable by a continuous feedback law.

---

[5] The original result was stated for C^1 maps, with subsequent relaxations in [95, 131].

Example 6.1 (*The nonholonomic integrator*) Consider the control system on $\mathbb{R}^3$ defined by

$$\begin{cases} \dot{x}_1 = u_1 \\ \dot{x}_2 = u_2 \\ \dot{x}_3 = x_2 u_1 - x_1 u_2, \end{cases} \tag{6.1}$$

also referred to as the "*Heisenberg system*" [11, Sect. 1.8]. Regarding the equilibrium $x^\star = 0 \in \mathbb{R}^3$ under $u^\star = 0 \in \mathbb{R}^2$, although the linearization of (6.1) around $(0, 0) \in \mathbb{R}^3 \times \mathbb{R}^2$ does not provide a controllable linear system, the system—conveniently written as $\dot{x} = g_1(x)u_1 + g_2(x)u_2$—is (globally) *controllable* as span $\{g_1, g_2, [g_1, g_2]\} = \mathbb{R}^3$ and (6.1) is drift-free [94, Chap. 3]). Here, $[\cdot, \cdot]$ denotes the *Lie bracket* [78]. Nevertheless, $(x, u) \mapsto g_1(x)u_1 + g_2(x)u_2$ is not surjective as one cannot map to $(0, 0, \varepsilon)$ for any $\varepsilon \neq 0$. See [2, Sect. 4] for a class of controllable systems that cannot be stabilized by smooth feedback and see [11, 19] for further generalizations of this example.

Observe that in Example 6.1 the topological obstruction is implicit in the dynamics, one cannot move freely in $\mathbb{R}^3$. Or as put by for example Sontag in [117], the nonholonomic system imposes "*virtual obstacles*". The fact that the linearization of (6.1) is not informative is inherent to nonholonomic systems. See [65, 123] for a principled methodology to obtain informative approximations using sub-Riemannian geometry. In fact, this viewpoint was exploited in the early work by Brockett on (6.1), cf. [17].

Moreover, note that Brockett's condition prevails when one allows for *dynamic feedback*. Also note that Brockett's condition essentially provides a necessary (but not sufficient) condition for $\mathrm{ind}_0(f) \neq 0$, while we know more. This leads to Zabczyk's index condition, where effectively one adapts Example 3.4 to the C^0 setting.

Theorem 6.2 (Zabczyk's index condition [131]) *Let $\Sigma_{n,m}^{\mathrm{loc}}$ be a local continuous control system. If a continuous feedback $x \mapsto \mu(x) \in \mathbb{R}^m$ with $\mu(0) = 0$ renders $0 \in \mathbb{R}^n$ locally asymptotically stable, then, the vector field index of $x \mapsto f(x, \mu(x))$ equals $(-1)^n$.*

Proof For flows this follows directly from Example 3.4. Momentarily ignoring completeness, due to asymptotic stability a smooth Lyapunov function must exist [49, 74, 126]. Following [34, p. 291], let V be a smooth Lyapunov function defined on some open ball $\mathbb{B}_\varepsilon^n(0)$ around 0. By construction one has $\langle \partial_x V(x), f(x, \mu(x)) \rangle < 0$ for all $x \in \mathbb{B}_\varepsilon^n(0) \setminus \{0\}$. Hence, $\partial_x V(x) \neq 0$ for all $x \in \mathbb{B}_\varepsilon^n(0) \setminus \{0\}$. In fact, V is Lyapunov function for $\dot{x} = -\partial_x V(x)$, hence $\mathrm{ind}_0(-\partial_x V) = (-1)^n$. Now we construct the map $H(s, x) = -s\partial_x V(x) + (1 - s)f(x, \mu(x))$, which is a vector field homotopy since $\langle H(s, x), \partial_x V(x) \rangle < 0$ for all $s \in [0, 1]$ and $x \in \mathbb{B}_\varepsilon^n(0) \setminus \{0\}$. Hence, $\mathrm{ind}_0(f) = \mathrm{ind}_0(-\partial_x V) = (-1)^n$.

Observe that the homotopy constructed in the proof above preserves stability along the homotopy since $\langle H(s, x), \partial_x V(x) \rangle < 0$ for all $s \in [0, 1]$. In other words, f can be transformed into the negated gradient flow $-\partial_x V = -\nabla V$ through qualitatively

Fig. 6.2 Given a nominal
vector field f' with
$\mathrm{ind}_0(f') = (-1)^n$. Then all
vector fields f that are
point-wise contained in
blunt, (salient) convex cones,
containing f', have the same
index as f'

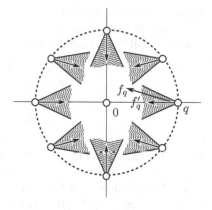

equivalent dynamical systems. Understanding when this can be done is of independent interest and has been studied in the context of gradient flows [104], locally [32, Chap. 9] and under convexity assumptions [67]. See Sect. 8 for open questions.

Although Zabczyk's condition is stronger than Brockett's condition, it is not necessary for dynamic feedback in the sense that the *"non-lifted"* dynamical control system $\dot{x} = f(x, u)$ might fail to satisfy the index condition while a stabilizing dynamic feedback exists [35]. The intuition is that the auxiliary state representing the input does not need to converge to 0 in an asymptotically stabilized manner. A secondary contribution of Zabczyk's work was to streamline arguments via bringing the work due Krasnosel'skiĭ and Zabreĭko to the attention, e.g., [73, Theorem 52.1]. An extension to constrained (unilateral) dynamical systems is presented in [51]. Now recall Example 3.2, then geometrically, Theorem 6.2 states that one needs to be able to find a continuous map μ such that the closed-loop vector field f "points in the direction" of a vector field f' with $\mathrm{ind}_0(f') = (-1)^n$. This is visualized in Fig. 6.2. Note, outward pointing cones would also work here as $(1)^{n=2} = (-1)^{n=2}$.

Zabczyk's condition is unfortunately not necessarily easy to check. Then, Coron's homological index condition captures the fact that the index should be ±1 via a condition on the vector field f, whose proof follows from the homological interpretation of the degree as sketched in Sect. 4.1, see also [59, Sect. 2.2]. Here we recall that f_* denotes the induced homomorphism.

Theorem 6.3 (Coron's condition [33, Theorem 2]) *Let $\Sigma_{n,m}^{\mathrm{loc}}$ be a local continuous control system with $n \geq 2$ and $m \geq 1$. Assume that f is continuous over some neighbourhood Ω of $(0, 0) \in \mathbb{R}^n \times \mathbb{R}^m$ and define $\Omega_\varepsilon = \{(x, u) \in \Omega : f(x, u) \neq 0, \|x\| < \varepsilon, \|u\| < \varepsilon\}$ with $\varepsilon \in \mathbb{R}_{>0} \cup \{+\infty\}$. If there is a continuous feedback $x \mapsto \mu(x) \in \mathbb{R}^m$ with $\mu(0) = 0$ rendering $0 \in \mathbb{R}^n$ locally asymptotically stable, then, for any $\varepsilon \in \mathbb{R}_{>0} \cup \{+\infty\}$*

$$f_*(H_{n-1}(\Omega_\varepsilon; \mathbb{Z})) = H_{n-1}(\mathbb{R}^n \setminus \{0\}; \mathbb{Z}) \quad (\simeq \mathbb{Z}). \tag{6.2}$$

Proof (*Sketch*) Assume that there is a continuous feedback rendering 0 locally asymptotically stable, say μ, with $\mu(0) = 0$. By continuity, there must be a sufficiently small $\delta > 0$ such that for all $x \in \mathrm{cl}\,\mathbb{B}_\delta^n(0) \setminus \{0\}$ one has $(x, \mu(x)) \in \Omega_\varepsilon$.

$$\mathrm{cl}\,\mathbb{B}_\delta^n(0) \setminus \{0\} \xrightarrow{\ g\ } \mathbb{R}^n \setminus \{0\}$$

Let $g(x) = f(x, \mu(x))$ and $v(x) = (x, \mu(x))$, then, the diagram above commutes. The map g represents the closed-loop vector field and as such, the vector field index of g, with respect to 0, over a sufficiently small open neighbourhood U of 0, must be $(-1)^n$. Differently put, we know that for sufficiently small δ, g is vector field homotopic to $-\mathrm{id}$ on $\mathrm{cl}\,\mathbb{B}_\delta^n(0) \setminus \{0\}$, see either Example 3.4, [34, Proof of Theorem 11.4] or [116, Proof of Theorem 22]. Indeed, one can also first homotope to a smooth vector field if desired, e.g., to $-\nabla V$. Now, recall (4.2), as $H_{n-1}(\mathrm{cl}\,\mathbb{B}_\delta^n(0) \setminus \{0\}; \mathbb{Z}) \simeq \mathbb{Z}$, the induced homomorphism $g_\star$ simply becomes

$$g_\star(H_{n-1}(\mathrm{cl}\,\mathbb{B}_\delta^n(0) \setminus \{0\}; \mathbb{Z})) = H_{n-1}(\mathbb{R}^n \setminus \{0\}; \mathbb{Z}) \quad (\simeq \mathbb{Z}).$$

Then the result follows by commutativity of the diagram, in particular the compositional property of the induced homomorphism(s). ∎

As highlighted in the proof sketch above, the condition $n \geq 2$ relates to the same condition in (4.2) ($H_0(\mathbb{S}^0; \mathbb{Z}) \not\simeq \mathbb{Z}$). Indeed, it suffices again to find an admissible continuous feedback law such that $x \mapsto f(x, \mu(x))$ has a $(-1)^n$ vector field index, cf. Theorem 6.2. Here, one must also check controllability or better yet *stabilizability*, otherwise see that $\dot{x} = x$ and $\dot{x} = -x$ both satisfy the aforementioned conditions (index ± 1), yet, the two systems have opposite qualitative properties. On $\mathbb{R}^{2n}$ both systems even have index $1 = (-1)^{2n}$. See [34, Exercise 11.7] for a typical example that satisfies Brockett's condition but fails to satisfy Coron's condition. As will be discussed in Sect. 7.2, time-varying feedback is frequently a solution. Generalizations of the index condition for *homogeneous* stabilization are presented in [114].

Surprisingly, Brockett's condition even prevails when certain forms of discontinuous feedback are allowed, as shown by Ryan [107]. Coron and Rosier showed a similar obstruction by concurrently linking the existence of a stabilizing discontinuous feedback to that of a stabilizing time-varying feedback [37], [36, Remark 7]. Discussing these obstructions in detail is outside the scope of this work as one needs to specify what *a solution* to a discontinuous differential equation (inclusion) means. Moreover, Ceragioli showed that this specification itself influences the obstructions [28].

Remark 6.1 (*Robustness/fragility*) Consider for some $\epsilon > 0$ the following perturbation of (6.1)

$$\begin{cases} \dot{x}_1 = u_1 \\ \dot{x}_2 = u_2 \\ \dot{x}_3 = \epsilon x_1 + x_2 u_1 - x_1 u_2. \end{cases} \tag{6.3}$$

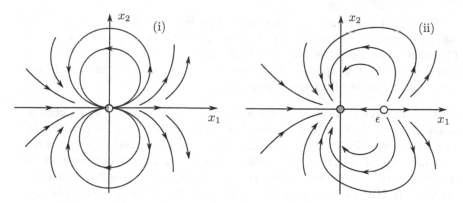

Fig. 6.3 Remark 6.1, (i): (6.4a) versus (ii): (6.4b) for $\epsilon = 1$

In contrast to (6.1), the linearization of this system, around $(0, 0) \in \mathbb{R}^3 \times \mathbb{R}^2$, *is* controllable from the origin for any $\epsilon > 0$. Yet, after linearizing (6.3) in a neighbourhood of $(0, 0)$, one finds that the smallest singular value of

$$\begin{bmatrix} 0 & 0 & 0 & 1 & 0 \\ 0 & 0 & 0 & 0 & 1 \\ \epsilon - u_2 & u_1 & 0 & x_2 & -x_1 \end{bmatrix},$$

subject to $u_1 = u_2 = 0$, is proportional to ϵ. Similarly, consider the *Artstein circles*

$$\begin{cases} \dot{x}_1 = x_1^2 - x_2^2 \\ \dot{x}_2 = 2x_1 x_2 \end{cases}. \tag{6.4a}$$

Here, $0 \in \mathbb{R}^2$ is globally attractive (if considered on $\mathbb{S}^2$), but not globally asymptotically stable and indeed, the corresponding vector field index is 2, as in Fig. 3.4(iv). Again, we perturb the system, this time such that 0 is *locally* asymptotically stable for any $\epsilon > 0$:

$$\begin{cases} \dot{x}_1 = -\epsilon x_1 + x_1^2 - x_2^2 \\ \dot{x}_2 = -\epsilon x_2 + 2x_1 x_2, \end{cases} \tag{6.4b}$$

see Fig. 6.3. Computing the linearization for (6.4b), one obtains the following matrix representation of the linear map

$$\begin{bmatrix} 2x_1 - \epsilon & -2x_2 \\ 2x_2 & 2x_1 - \epsilon \end{bmatrix}.$$

Therefore, around $0 \in \mathbb{R}^2$, the ϵ-perturbation turned 0 into a hyperbolic equilibrium point. Yet, in a neighbourhood around 0, the smallest singular (eigen) values are proportional to ϵ.

These examples relate to the fact that hyperbolic equilibrium points are generic and have a vector field index of ± 1. Similarly, recall Thom's transversality theorem (Theorem 3.1) and consider the genericity of smooth maps $f : \mathbb{R}^n \times \mathbb{R}^m \to \mathbb{R}^n$ such that $f \pitchfork \{0\}$. As such, applying the aforementioned conditions to models obtained, for example, numerically, requires great care.

Also, the original version of Theorem 6.1 [18, Theorem 1.(iii)] by Brockett was stated under a C^1 assumption. Although we only present the third condition, the first condition of [18, Theorem 1] further illuminates the interest in *continuous* feedback laws. Specifically, this condition states that the linearized system must not have unstable uncontrollable modes, not to obstruct the existence of a *differentiable* feedback cf. the Hartman–Grobman theorem [106, Chap. 5]. This condition, however, does not rule out the existence of continuous feedback laws, e.g., in [103] Qian and Lin exploit homogeneity and a cascadic structure to construct continuous controllers for systems that might not admit a smooth controller. See also the earlier work by Aeyels [2], Kawski [70] and Dayawansa, Martin and Knowles [40].

Example 6.2 (*Smooth versus continuous feedback*) Consider [69, Example 2], that is, the control system on $\mathbb{R}^2$ defined by

$$\begin{cases} \dot{x}_1 = 4x_1 + u x_2^2 \\ \dot{x}_2 = x_2 + u, \end{cases} \tag{6.5}$$

with $u \in \mathbb{R}$. The control system can be shown to be controllable, yet, no smooth, asymptotically stabilizing feedback $x \mapsto \mu(x) \in \mathbb{R}$ exists as the differential of (6.5) around 0 will have an unstable mode. However, (6.5) *does* satisfy Brockett's condition and also Coron's condition. After a feedback transformation, a non-differentiable change of coordinates and a reparametrization of time, (6.5) becomes

$$\begin{cases} \dot{z}_1 = z_1 - z_2^3 \\ \dot{z}_2 = v, \end{cases} \tag{6.6}$$

for the new input term $v \in \mathbb{R}$. Then, the results due to Kawski provide for an explicit continuous stabilizing feedback [70] for the control system (6.6).

6.1.2 Global Obstructions

Now we have the machinery to present the first illustrative global topological obstruction with respect to the domain of attraction.

Theorem 6.4 (Sontag's condition [116, Theorem 21]) *Let $p^\star$ be a locally asymptotically stable equilibrium point of $X \in \mathfrak{X}^0(M)$. Then, the set $\mathcal{D}(\varphi_X, p^\star)$ as given by (5.3) is open and contractible to $p^\star$.*

Proof (*Sketch*) Openness follows from continuity of φ_X^t. Regarding the contractibility, the natural candidate for the homotopy would be the flow with time being rescaled from $\mathbb{R}_{\geq 0}$ to $[0, 1]$, i.e., $H(t, p) = \varphi_X(t/(1 - t), p)$ with $t \in [0, 1)$ and a well-defined limit. For the full proof see [116, Theorem 2] or the remarks below Theorem 6.12.

Theorem 6.4 indicates that when M is not contractible, there is no $p^* \in M$ that happens to be globally asymptotically stable under the flow of some complete vector field $X \in \mathfrak{X}^r(M)$. Indeed, $\mathbb{R}^n$ *is* contractible and global asymptotic stabilization of an equilibrium is not immediately obstructed on such a space. It is instrumental to remark that *asymptotic* stability is exploited in Theorem 6.4. Moreover, we like to recall an example due to Takens [119, p. 231], responding to a question of Thom, showing that there is a (polynomial) gradient vector field X such that the topology of the set $\Gamma = \{p \in M : \lim_{t \to \infty} \varphi_X^t(x) = p^*\}$, for *some* equilibrium point $p^* \in M$, is not necessarily invariant under a change of Riemannian metric (intuitively, a change of coordinates). Indeed, the equilibrium point considered is not asymptotically stable.

Akin to (5.3) one defines $\mathcal{D}(\varphi_X, A)$ for A a compact attractor, cf. Chapter 5. Early documented results on the topology of attractors and their domain of attraction can be found in [10]. For example, in the Euclidean setting, $A \subset \mathbb{R}^n$ can only be a *globally* asymptotically stable compact set if $\mathbb{R}^n \setminus A$ is homeomorphic to $\mathbb{R}^n \setminus \{0\}$ [10, Theorem V 3.6]. Moreover, if $p^* \in \mathbb{R}^n$ is globally asymptotically stable under some flow φ, then $\mathcal{D}(\varphi, p^*) \setminus \{p^*\}$ is homeomorphic to $\mathbb{R}^n \setminus \{0\}$ [10, Corollary V 3.5]. Also, when A is a compact attractor, we have that A necessarily consists out of finite components [10, Theorem V 1.22]. Now, as homotopies preserve path-connected components, when $A \subset M$ consists out of more path-connected components than M, there is no continuous vector field on M with A globally asymptotically stable. Interestingly, globally asymptotically stabilizing a disconnected set cannot be achieved using robust hybrid feedback either [111]. Although outside the scope of this work, see [50] for related results in the infinite-dimensional setting.

Theorem 6.4 immediately implies that vector fields over non-contractible spaces do not admit globally asymptotically stable equilibrium points. This observation clarifies the obstructions we recovered for the circle $\mathbb{S}^1$ as shown in Figure 1.1. Moreover, as contractible spaces M are homotopy equivalent to a point they have (singular homological) Euler characteristic $\chi(M) = 1$, e.g., see (4.4).

Example 6.3 (*Case study Sect. 1.3: global stabilization is obstructed*) As was shown in Example 3.5, all compact Lie groups G have $\chi(G) = 0$, hence they are not contractible and global asymptotic stabilization by means of continuous feedback is prohibited. See Example 6.10 for non-compact examples.

As was the motivation for the case study, Example 6.3 is of importance in many dynamical systems grounded in mechanics as they can be frequently identified with Lie groups [5, 11, 22, 93, 113].

An immediate but appealing manifestation of this line of reasoning is the result by
Bhat and Bernstein [9]. As with Brockett's condition, their result contradicted what
was thought to be true at the time (second half of the twentieth century), cf. [90,
108].

Theorem 6.5 (Bhat–Bernstein condition [9, Theorem 1]) *Let* $\pi : \mathsf{M} \to \mathsf{B}^b$ *be a
vector bundle for some smooth, compact, boundaryless, base manifold* B^b *with* $b \geq 1$,
then, there is no continuous vector field on M *with an isolated globally asymptotically
stable equilibrium point.*

Proof We know from Lemma 3.1 that compact manifolds of the form B^b are never
contractible. From Example 3.1 we know that B^b is a deformation retract of M so that
M itself can also not be contractible by homotopy equivalence. Then, an application
of Theorem 6.4 concludes the proof.

The compactness of the base manifold can be relaxed to not being contractible and
the vector bundle can be generalized to a fiber bundle.[6] Theorem 6.5 is especially
of use in *second order* dynamical systems, e.g., Lagrangian systems in robotics
are frequently defined over compact manifolds [93]. Theorem 6.5 clearly obstructs
dynamic feedback,[7] and as was shown by Bernuau, Perruquetti and Moulay, also
uniform stabilization by means of time-varying feedback [8].

Example 6.4 (*Obstruction to time-varying stabilization* [8]) Theorem 6.5 obstructs
the existence of continuous dynamic feedback to globally asymptotically stabilize
an isolated equilibrium point over a fiber bundle. The reason being that in the case of
dynamic feedback, the augmented system is usually rendered asymptotically stable
as a whole, e.g., when employing an observer. Now if we consider time-varying feedback e.g., $\dot{s} = 1$, $\dot{x} = f(x, \mu(s, x))$, then, any solution will have its last component
diverge to $+\infty$ for $t \to +\infty$, hence the previous argument breaks down. To study
the situation we need to define *partial* stability. Let $\mathsf{N} = \mathsf{S} \times \mathsf{M}$ be a smooth product
manifold and let $X = (X_s, X_p)$ be a forward complete continuous vector field on
N, giving rise to the semiflow ϕ_X, that is, the domain of ϕ_X is $\mathbb{R}_{\geq 0} \times \mathsf{N}$ instead of
$\mathbb{R} \times \mathsf{N}$, and let π_2 be the projection from N onto M. Note that $X_s : \mathsf{S} \times \mathsf{M} \to T\mathsf{S}$
captures more time-varying schemes than simply $\dot{s} = 1$. Now, the point $x_\infty \in \mathsf{M}$ is
said to be a *partial equilibrium* of X when $X_p(s, x_\infty) = 0$ for all $s \in \mathsf{S}$. Then, x_∞
is partially stable uniformly in s when for any neighbourhood $U_\epsilon \subseteq \mathsf{M}$ of x_∞ there
is a neighbourhood $U_\delta \subseteq \mathsf{M}$ of x_∞ such that for all $x \in U_\delta$ and for all $s \in \mathsf{S}$ one has
$\pi_2 \circ \phi_X(t, (s, x)) \in U_\epsilon$ for all $t \geq 0$. Now x_∞ is *partially globally asymptotically
stable uniformly in* s when it is partially stable uniformly in s and for all $(s, x) \in \mathsf{N}$

---

[6] Here, one should assume that the fiber bundle admits a (global) section for the statement not to be
vacuous, e.g., by having contractible fibers. For the prototypical counter example remove $Z_\pi(\mathbb{S}^2)$
from $T\mathbb{S}^2$ or let $\mathbb{S}^1 \hookrightarrow \mathbb{C}$ be the base manifold with the fiber $\mathbb{Z}_2 = \mathbb{Z}/2\mathbb{Z}$, i.e., one has the map
$\pi : \mathbb{S}^1 \to \mathbb{S}^1$ defined by $\pi : z \mapsto z^2$. However, the map $\sigma : z \mapsto z^{1/2}$ fails to be continuous on
$\mathbb{S}^1 \hookrightarrow \mathbb{C}$.

[7] Theorem 6.5 boils down to stabilizing feedback being obstructed due to M not being contractible.
A common approach to dynamic feedback would enlarge the state space, e.g., to $\mathsf{M} \times \mathsf{U}$, which can
never be contractible if M is not.

one has $\pi_2 \circ \phi_X(t, (s, x)) \to x_\infty$ for $t \to +\infty$. Regarding *partial global asymptotic stability uniform in s*, consider for example $\dot{x} = -xs$, $\dot{s} = 1$. To continue, let $\pi : M \to Q$ be a vector bundle with Q a C^∞ compact base manifold without boundary and assume that a stabilizing time-varying feedback does exist, that is, there is a continuous vector field $X = (X_s, X_p)$ over $N = S \times M$ with some point $x_\infty \in M$ being partially globally asymptotically stable uniformly in s. Let $q_\infty = \pi(x_\infty)$ and define $\sigma(q) = (s', \sigma'(q))$ for some section $\sigma' \in \Gamma^0(M)$ and some $s' \in S$. It follows that σ is a section of $\pi \circ \pi_2 : N \to Q$. By assumption, X gives rise to a semiflow ϕ_X, then consider the map $H : [0, 1] \times Q \to Q$ defined by

$$(\lambda, q) \mapsto \begin{cases} \pi \circ \pi_2 \circ \phi_X \left(\log\left(1/(1 - \lambda)\right), \sigma(q) \right) & \text{if } \lambda \neq 1 \\ q_\infty & \text{if } \lambda = 1. \end{cases}$$

Indeed, $H(0, q) = q$ and $H(1, q) = q_\infty$. However, as one can show that H is continuous, this homotopy contradicts Q being closed. Therefore, such a time-varying feedback cannot exist. Again, one can generalize the construction to fiber bundles under the assumption that a continuous section exists. All details can be found in [8] and see for instance [34, Sect. 11.2] for more on time-varying stabilization.

Furthermore, as mentioned in [9], Theorem 6.5 can be applicable to *non*-compact configuration spaces, as long as M can be written as a (vector) bundle, e.g., let Q^q be compact with a trivial tangent bundle, like $\mathbb{S}^1$, then the configuration space $Q^q \times \mathbb{R}^p$ leads to $M = T(Q^q \times \mathbb{R}^p) \simeq Q^q \times \mathbb{R}^{q+2p}$.

One of the objectives of this work is to clarify to what extent Theorem 6.5 prevails when the *global* stability condition is relaxed to *local multistability* conditions, or when we look beyond the stabilization of points. As such, we first consider a variety of necessary conditions which indicate if a (compact) manifold admits a continuous vector field with all of its zeroes being isolated locally asymptotically stable equilibrium points.

Theorem 6.6 (A global necessary condition for local stability) *Let* M^n *be a smooth, compact, boundaryless, finite-dimensional manifold. Then,* M^n *admits a continuous vector field* $X \in \mathfrak{X}^{r \geq 0}(M^n)$ *with* $z \in \mathbb{N}_{\geq 0}$ *zeroes, all of which are isolated and locally asymptotically stable, if and only if* $\chi(M^n) = z$.

Proof See that the special case of $z = 0$ is handled by Proposition 3.6. Then, first for the "*only if*" direction. Consider the case $z > 0$, as the index of these locally asymptotically stable equilibrium points is $(-1)^n$ (see Example 3.4), we see that $\chi(M^n)$ must equal $z(-1)^n$. By Theorem 6.8, this cannot be true for odd-dimensional manifolds. Therefore, a necessary condition for local asymptotic multistability is that $\chi(M^n) = z$. For the "*if*" direction one can follow the same line of arguments as used to show Proposition 3.6, e.g., see [54, pp. 141–148]. The difference being that now one starts, in local coordinates, from a set of locally asymptotically stable equilibrium points such that sum of their indices equals $\chi(M^n)$. The equilibrium points that emerge when these local vector fields are patched together on M^n are removed precisely as in a general proof of Proposition 3.6.

Theorem 6.6 generalizes [91, Theorem 25] and the necessary direction can also be shown along the lines of [72] or [87, Chap. 15], see also [109, Example 17].

The next example clarifies how compact manifolds M with $\chi(M) = 1$ do not contradict the theory, *global* asymptotic stability remains obstructed.

Example 6.5 (*Grassmannian manifolds*)   Let $Gr(k, n)$ denote the set of k-dimensional subspaces of $\mathbb{R}^n$, which is a smooth, compact, boundaryless, $k(n - k)$-dimensional, non-orientable manifold [1, Sect. 3.4.4]. These *Grassmannian* manifolds appear frequently in the context of manifold optimization [1, 14]. Interestingly, one can compute that $\chi(Gr(2, 3)) = 1$, e.g., by considering the normal to a subspace one can identify ($\simeq_t$) $Gr(2, 3)$ with the real projective plane $\mathbb{RP}^2$, defined as $\mathbb{RP}^2 = (\mathbb{R}^3 \setminus \{0\})/ \sim$ with $p \sim \lambda p$ for any $\lambda \in \mathbb{R} \setminus \{0\}$, and consider a 2-sheeted covering of $\mathbb{RP}^2$ by $\mathbb{S}^2$.[8] Indeed, this implies that there is a vector field on $Gr(2, 3)$ with a single locally asymptotically stable equilibrium point. However, as $Gr(2, 3)$ is compact, it is not contractible, hence the point cannot be *globally* asymptotically stable or most of Chap. 3 would be contradicted. Indeed, again identify $Gr(2, 3)$ with $\mathbb{RP}^2 = \mathbb{S}^2/ \sim$, for $p \sim q$ when $p = -q$. Then consider the vector field as displayed in Fig. 6.4 and observe the periodic orbit on the equator. The resulting vector field on $\mathbb{RP}^2$, or $Gr(2, 3)$ for that matter, will have a non-trivial limit set.[9] This observation is further discussed in Sect. 8.4.

The previous example is of mathematical interest, but interestingly, the human eye (gazing) can be modelled as a control system on a space diffeomorphic to $\mathbb{RP}^2$ [101]. Then, with respect to Example 6.5, we leave it to the reader to infer potential physical ramifications, e.g., does the non-trivial (α)-limit set relate to *nystagmus* (unintentional oscillatory movement of the eye)?

In line with the work on almost global stability, Theorem 6.6 provides motivation for abandoning local asymptotic multistability as many spaces have their Euler characteristic being equal to 0. A straightforward necessary conditions follows.

Corollary 6.1 (A necessary condition for local asymptotic multistability) *Any, smooth, compact, boundaryless, finite-dimensional manifold* M *admits a continuous vector field* $X \in \mathfrak{X}^{r \geq 0}(M)$ *with all of its* $z \in \mathbb{N}_{>0}$ *zeroes being locally asymptotically stable isolated equilibrium points only if* $\chi(M) > 0$.

---

[8] Returning to Chap. 3, we remark how the Poincaré–Hopf theorem remains true when the underlying manifold is not orientable. Given two smooth manifolds M' and M, a map $\pi' : M' \to M$ is called a *smooth covering* when π' is a smooth surjective map and for each $p \in M$ there is neighbourhood U of p such that each *component* of $\pi'^{-1}(U)$ is mapped diffeomorphically onto U [78, p. 91]. Now, define the manifold $\widehat{M} = \{(p, O_p) : p \in M, O_p \text{ an orientation on } T_pM\}$. and the map $\widehat{\pi} : \widehat{M} \to M$ by $\widehat{\pi}(p, O_p) = p$. Such a map is called the *orientation covering* of M. Let $\widehat{\pi} : \widehat{M} \to M$ be such an orientation covering. Then, if M is a connected smooth non-orientable manifold, $\widehat{M}$ is a connected smooth oriented manifold and $\widehat{\pi}$ is a *two*-sheeted covering [78, pp. 392–396]. This also motivates the mod *two* construction. It follows that if N is a k-sheeted cover of M, then, $\chi(N) = k\chi(M)$. Now, given a vector field X on M, using $\widehat{\pi} : \widehat{M} \to M$, pull X back to a vector field on $\widehat{M}$, which can be done as $\widehat{\pi}$ is a component-wise diffeomorphism, then, apply the Poincaré–Hopf theorem and divide by 2.

[9] As $\mathbb{RP}^2$ does not embed in $\mathbb{R}^3$, this can, however, not be easily correctly visualized.

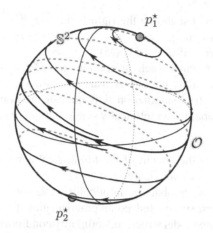

Fig. 6.4 In line with $\chi(\mathbb{S}^2) = 2$ and Theorem 6.6, a vector field on $\mathbb{S}^2$ with two isolated locally asymptotically stable equilibrium points $(p_1^\star, p_2^\star)$ and an unstable periodic orbit O. This vector field also exemplifies the Poincaré–Bendixson theorem, locally, e.g., see [98, Theorem 1.8]

As for manifolds like any odd-sphere $\mathbb{S}^{2n+1}$, the n-torus and any Lie group, $\chi(M) = 0$. Hence, for many spaces, local asymptotic multistability is impossible. Regarding a negative Euler characteristic, recall that closed orientable surfaces M^2 with genus g have $\chi(M^2) = 2 - 2g$.

Remark 6.2 (*Theorem 6.4 compared with Theorem 6.6*) Consider any even n-sphere like $\mathbb{S}^2$, using Sontag's condition one concludes that continuous global asymptotic stability is impossible on such a topological manifold. Relaxing the adjective *global*, Fig. 6.4 shows a locally asymptotically multistable vector field on precisely $\mathbb{S}^2$. Indeed, the necessary condition from Theorem 6.6 holds true as for this example $\chi(\mathbb{S}^2) = 2$, which equals the number of isolated equilibrium points. Note that in the case of Fig. 6.4, the equator functions as an unstable equilibrium point, but by means of an unstable periodic orbit, not a point. Nevertheless, as most relevant compact manifolds M have $\chi(M) = 0$, Theorem 6.4 extends to multistability in the sense that for most compact manifolds local asymptotic multistability is prohibited.

Example 6.6 (*Opinion dynamics*) In [6], the authors study *opinion dynamics* on the compact manifolds $\mathbb{S}^1$, $\mathbb{S}^2$ and $\mathbb{T}^2$ with the purpose of understanding how the underlying state space influences the formation of opinions. Omitting a few details, the authors consider

$$\frac{\mathrm{d}}{\mathrm{d}t}x_i(t) = \sum_{j=1}^{N} a_{ij} \Psi\left(d(x_i(t), x_j(t))\right) v_{ij} \tag{6.7}$$

for $x = (x_1, x_2, \ldots, x_N) \in \times_{i=1}^{N} M = (M)^N$ denoting the set of opinions, $a_{ij} \in \mathbb{R}$ the interaction coefficients, $\Psi : \mathbb{R} \to \mathbb{R}$ the interaction potential with $\Psi(0) = 0$, $d : M \times M \to \mathbb{R}_{\geq 0}$ a (Riemannian) distance on M and $v_{ij} \in T_{x_i}M$ the direction of influence.

Assume that Ψ is selected such that the right-hand-side of (6.7) is continuous. As expected, the *consensus* setting $x_1 = x_2 = \cdots = x_N$ is a set of equilibrium points: the *consensus manifold*, denoted $C = \{x \in (M)^N : x_1 = x_2 = \cdots = x^N\}$. Is consensus the the only equilibrium and is such an opinion stable? As $\chi((M)^N) = N \cdot \chi(M)$, manifolds like $\mathbb{S}^2$ cannot only have consensus as equilibria. In fact, the upcoming obstructions to submanifold stabiliztion will indicate that on compact manifolds consensus is never a globally asymptotically stable stationary opinion when M is not contractible.

Example 6.6 exemplifies the benefit of the topological approach, we can arrive at strong insights without any explicit study of (6.7).

Summarizing the above, global asymptotically stable equilibrium points are rare to encounter on nonlinear spaces. Indeed, in practice, global notions of stability rely on for instance exploiting model structure, optimality conditions and the existence of Lyapunov-like functions, e.g., in [124], the authors consider Jurdjevic–Quinn type systems. Chap. 7 presents more methods towards (almost) global stability.

6.1.3 A Local Odd-Number Obstruction to Multistabilization

In this section multistability is considered locally. In particular, the results are in the spirit of the work by Ortega [96] and closely related to the *odd-number limitation* in delayed feedback, which we briefly introduce below. To aid the exposition, we momentarily deviate from the continuous-time system (5.1) and start with a linear discrete-time system.

Following [129] we introduce the so-called *odd-number limitation*, which naturally appears in the context of *delayed* feedback control, as pioneered by Pyragas [102]. Consider a deterministic linear time-invariant system

$$\begin{cases} x_{k+1} = Ax_k + Bu_k, \\ y_k = Cx_k, \end{cases} \tag{6.8}$$

with $x \in \mathbb{R}^n$ representing again the state, $u \in \mathbb{R}^m$ the input and $y \in \mathbb{R}^p$ the output (the observables), for some real matrices A, B and C of appropriate size. Now assume we want to control (6.8) using the delayed linear feedback $u_k = K(y_k - y_{k-1})$ for some matrix K, as is one of the key methods in the control of periodic orbits. The resulting closed-loop system can be written as

$$z_{k+1} = (A' + B'KC')z_k \tag{6.9}$$

for

$$A' = \begin{bmatrix} A & 0 \\ I_n & 0 \end{bmatrix}, \quad B' = \begin{bmatrix} B \\ 0 \end{bmatrix}, \quad C' = \begin{bmatrix} C & -C \end{bmatrix}, \quad z_k = \begin{bmatrix} x_k \\ x_{k-1} \end{bmatrix}.$$

Let $P : \mathbb{C} \to \mathbb{C}$ be the characteristic polynomial of the closed-loop system defined by $P(\lambda) = \det(\lambda I_n - A' - B'KC')$ with $\lambda = a + ib$, then, for (6.9) to be asymptotically stable, we need[10] $P(\lambda) \neq 0$ for all $|\lambda| \geq 1$. As A' is even-dimensional $\lim_{a \to +\infty} P(|a| + i0) = +\infty$ irrespective of our definition of $P(\lambda)$ i.e., $\det(\lambda I_n - (\cdot))$ versus $\det((\cdot) - \lambda I_n)$. Then, asymptotic stability of (6.9) together with continuity of the determinant implies that $P(1 + i0) > 0$. This implies, by exploiting the block structure[11] of the problem, that $\det(I_n - A) > 0$ must hold for (6.9) to be asymptotically stable. Hence we observe—independent of the choice of K—a manifestation of the odd-number limitation, that is, unstable eigenvalues with $\lambda_i(A) > 1$ must come in pairs to allow for stabilization.

For more on this phenomenon, see [3, 62]. The odd-number limitation extends to more involved settings, and as recently shown, to hyperbolic equilibrium points [42]. Next, we go back to the continuous-time setting and highlight an obstruction to *Pyragas control*, e.g., stabilization of τ-periodic orbits by means of τ-delayed feedback.

Theorem 6.7 (De Wolff and Schneider [42, Corollary 3]) *Consider the differentiable control system $\dot{x}(t) = f(x(t)) + u$ over $\mathbb{R}^n$ and suppose that $p^\star \in \mathbb{R}^n$ is a non-degenerate equilibrium of f. Moreover, assume that $Df_{p^\star}$ has an odd number of eigenvalues (counting algebraic multiplicities) in the strict right half (complex) plane. Then, for all $K \in \mathbb{R}^{n \times n}$ and all $\tau > 0$, $p^\star$ is unstable as a solution of the controlled system $\dot{x}(t) = f(x(t)) + K[x(t) - x(t - \tau)]$.*

In the setting of compact manifolds, Theorem 6.7 extends, under more general types of feedback. This condition is useful in situations where $\chi(M)$ is unavailable and/or one has no knowledge of the number of equilibrium points of $X \in \mathfrak{X}^r(M)$, but one does have some local information.

Theorem 6.8 (Odd-number obstruction to local asymptotic multistabilization) *Let $X \in \mathfrak{X}^{r \geq 0}(M^n)$ be an uncontrolled vector field on a smooth, compact, boundaryless manifold M^n with all its zeros being isolated hyperbolic equilibrium points $\{p_1^\star, \ldots, p_{|I|}^\star\} \neq \emptyset$. Given any control system $\Sigma = (M^n, \mathcal{U}, F)$ in the sense of Definition 5.5, then, the set $\{p_i^\star\}_{i \in I}$ can be locally asymptotically stabilized by continuous feedback $\mu : M^n \to \mathcal{U}$, without introducing new equilibrium points, only if $\dim(W^u(\varphi_X, p_i^\star))$ is even for all $i \in I$.*

Proof Without loss of generality, we will consider some $I \neq 0$ and an even-dimensional manifold as the odd case cannot be handled regardless, e.g., recall Corollary 4.1. If $\dim(W^u(\varphi_X, p_i^\star))$ is odd, then $\mathrm{ind}_{p_i^\star}(X) = -1$ and as such, by the Poincaré–Hopf theorem (Corollary 3.1) and the hyperbolic index results from [73, Secti. 6], $|I| \neq \chi(M^n)$. Hence, by Theorem 6.6, this set cannot be locally asymptotically stabilized by means of continuous state-feedback.

---

[10] Asymptotic stability of linear discrete-time systems of the form $x_{k+1} = Ax_k$ is captured by $\rho(A) < 1$, for $\rho(A)$ the spectral radius of A. Such a A is also said to be *Schur stable*.

[11] Let some square matrix D be invertible, then a useful decomposition for appropriately sized square matrices A, B, C is the following $\begin{bmatrix} A & B \\ C & D \end{bmatrix} = \begin{bmatrix} I & BD^{-1} \\ 0 & I \end{bmatrix} \begin{bmatrix} A - BD^{-1}C & 0 \\ 0 & D \end{bmatrix} \begin{bmatrix} I & 0 \\ D^{-1}C & I \end{bmatrix}$.

Instead of demanding that $\dim(W^u(\varphi_X, p_i^*))$ is not odd, one could equivalently demand that, in coordinates, the differential of the uncontrolled vector field X satisfies $DX_{p_i^*} \in \mathsf{GL}^+(n, \mathbb{R})$ for all $i \in I$. As the orientation of a map is a *topological invariant* [77, Chap. 6], this implies that the statement of Theorem 6.8 is topologically invariant, as it should be. For example, let a^* and b^* correspond to two hyperbolic equilibrium points of some vector fields X and Y, respectively, with the dynamics around those equilibria locally captured by $\dot{x} = Ax$ and $\dot{y} = By$ for a^* and b^*, respectively. Then, since $\mathsf{GL}(n, \mathbb{R}) = \mathsf{GL}^+(n, \mathbb{R}) \sqcup \mathsf{GL}^-(n, \mathbb{R})$ only if the maps $f(x) = Ax$ and $g(y) = By$ have the same orientation, there is a homotopy $H(s, z) = C(s)z$ such that $H(0, x) = Ax$ and $H(1, y) = By$ with $\dot{z} = C(s)z$ hyperbolic for all $s \in [0, 1]$ [106]. Better yet, let $A \in \mathbb{R}^{n \times n}$ correspond to an asymptotically stable linear system, i.e., $\mathrm{ind}_0(Ax) = (-1)^n$ then for $\mathrm{ind}_0(By)$, with $B \in \mathbb{R}^{n \times n}$, to be $(-1)^n$, one needs the unstable eigenspace of B to be of even dimension. In other words, the even dimension is necessary for local analysis. This is not an overly strong condition as spirals and second order systems are naturally of even dimension, see also the discussions in [42]. Note that although $\mathsf{GL}^+(n, \mathbb{R})$ is a subgroup of $\mathsf{GL}(n, \mathbb{R})$, it is not necessary in general to preserve the group structure as (locally) asymptotically stable systems are not necessarily hyperbolic, e.g., consider $\dot{x} = -x^3$.

Example 6.7 (*Rayleigh quotient on the sphere*) Consider a smooth function $\ell : \mathbb{S}^{n-1} \subset \mathbb{R}^n \to \mathbb{R}$ defined by $\ell : p \mapsto \frac{1}{2}\langle Ap, p \rangle$ for some symmetric matrix $A \in \mathsf{Sym}(n, \mathbb{R})$ and $\langle \cdot, \cdot \rangle$ the Euclidean inner-product. One might be interested in either minimizing or maximizing ℓ over $\mathbb{S}^{n-1}$. By means of this function we will study gradient flows on $\mathbb{S}^{n-1} \subset \mathbb{R}^n$ and exemplify previously discussed material. When one views $\mathbb{S}^{n-1}$ as a Riemannian submanifold of $\mathbb{R}^n$, then $\mathrm{grad}\,\ell(p) = (A - I_n\langle Ap, p \rangle)p$ for all $p \in \mathbb{S}^{n-1}$ [1, Table 4.1].

(i) Regarding Theorem 6.6, $v_i \in \mathbb{S}^{n-1}$ is a unit eigenvector of A if and only if v_i is a critical point of the function ℓ [1, Proposition 4.6.1]. As $\chi(\mathbb{S}^{n-1}) = 0$ for n being even and $\chi(\mathbb{S}^{n-1}) = 2$ for n being odd, $\chi(\mathbb{S}^{n-1}) \neq n$ for all $n \in \mathbb{N}_{\geq 0}$. Hence, by Theorem 6.6 there is no continuous vector field $X \in \mathscr{X}^{r \geq 0}(\mathbb{S}^{n-1})$ such that every equilibrium point of X is isolated and locally asymptotically stable but also an eigenvector of $A \in \mathsf{Sym}(n, \mathbb{R})$.

(ii) To exemplify Theorem 6.8, let $n = 3$ and assume to be an observer of the gradient flow (under the Euclidean metric) $\mathrm{grad}\,\ell$ (maximization) on $\mathbb{S}^2 \subset \mathbb{R}^3$, without having knowledge of ℓ. One can show that this flow is equivalent to the projection of the solution to $\dot{x}(t) = Ax(t)$ onto the sphere. Let A have only simple eigenpairs (v_i, λ_i) with $\lambda_1 < \lambda_2 < \lambda_3$, it follows from [1, Proposition 4.6.2] that for a curve $t \mapsto \gamma(t)(v_j + tv_i)/\|v_j + tv_i\|_2$ with $i \neq j$ one has

$$\frac{\mathrm{d}^2}{\mathrm{d}t^2}\ell(\gamma(t))|_{t=0} = \lambda_i - \lambda_j.$$

This implies that only the points $\pm v_3 \in \mathbb{S}^2$ are locally asymptotically stable equilibrium points while $\pm v_1$ and $\pm v_2$ are unstable. Indeed, the unstable manifold of $\pm v_2$ is odd-dimensional, obstructing the possibility of global continuous multistabilization as set forth by Theorem 6.8. See also [60, Sect. 1.3].

(iii) At last, recall Brockett's necessary condition, i.e., Theorem 6.1, and consider a Rayleigh-like control system on $\mathbb{S}^2 \subset \mathbb{R}^3$ defined by

$$\dot{x} = f(x, u) = (A - I_3 \langle Ax, x \rangle)x + (I_3 - xx^\mathsf{T})u \qquad (6.10)$$

for some symmetric matrix $A \in \mathsf{Sym}(3, \mathbb{R})$ and $u \in \mathbb{R}^3$. As $T_x \mathbb{S}^2 = \{(I_3 - xx^\mathsf{T})v : v \in \mathbb{R}^3\}$, Brockett's condition holds, locally, but we know from above that (continuous) local asymptotic multistability cannot hold for (6.10). Hence, the local condition proposed in [18] is too weak to generalize to multistable problems on compact manifolds (the global topology is not taken into account). In contrast, Theorem 6.8 catches the impossibility correctly.

See [120, Corollary 5] for an odd-number obstruction in the context of network control and [30, Theorem 1] for odd-number results in the context of optimization. See also [115, Proposition 4.11] for a result technically in the spirit of Theorem 6.8, but with ramifications in the study of *Anosov diffeomorphisms*, i.e., diffeomorphisms $G : \mathsf{M} \to \mathsf{M}$ such that $T\mathsf{M}$ has a hyperbolic structure under G.

6.2 Obstructions to the Stabilization of Submanifolds

Consider again the nonholomic integrator (6.1). We know that the origin cannot be locally asymptotically stabilized by continuous feedback, but what about another set? If one ignores the x_3 coordinate we expect to be able to stabilize sets in some $x_1 - x_2$ plane. Indeed, simply pick $u_1 = -x_1$ and $u_2 = -x_2$ to locally asymptotically stabilize $(0, 0, x_3(t_0)) \in \mathbb{R}^3$. This intuition extends and one can for example find controllers to locally asymptotically stabilize the cylinder $\{x \in \mathbb{R}^3 : x_1^2 + x_2^2 = 1\}$ [82, p. 1], see also [89, Example 5]. As stated by Mansouri, when stabilization to a point fails, stabilization to a submanifold seems the next best. Submanifold stabilization occurs for example naturally in the context of nonholonomic control systems [12] and feedback linearization [64, 94]. In particular, if stabilization to a point fails, can one stabilize the system to a small Euclidean sphere enclosing this point?

To start, let A be a compact, connected, oriented, codimension-1 embedded submanifold of $\mathbb{R}^n$, i.e., $\dim(\mathsf{A}) = n - 1$, and let d denote the Euclidean metric.[12] Using ideas similar to Coron [33], Mansouri proves the following.[13] Again, we recall that $f_\star$ denotes the induced homomorphism from Chap. 4.

Theorem 6.9 (Mansouri's condition [82, Proposition 1]) *Let* $\Sigma_{n,m}^{\mathrm{loc}}$ *be a local continuous control system. Assume that f is continuous over some open neighbourhood* Ω

---

[12] The distance from $p \in \mathbb{R}^n$ to a compact set $\mathsf{A} \subset \mathbb{R}^n$ is defined as $d(p, \mathsf{A}) = \min_{a \in \mathsf{A}} \|p - a\|$.
[13] Here we take the recent results due to Fathi and Pageault into account [49], cf. [83, Remark 2.2]. See also [79].

of $\subset \mathbb{R}^n \times \mathbb{R}^m$ *and define* $\Omega_{f,\varepsilon} = \{(x, u) \in \Omega : f(x, u) \neq 0, d(x, \mathsf{A}) < \varepsilon\}$. *Let* $x \mapsto$ $\mu(x)$ *be a continuous feedback rendering* A *an attractor with* $g(x) = f(x, \mu(x))$ *denoting the closed-loop. Then, one has for all* $\varepsilon > 0$

$$f_\star(H_{n-1}(\Omega_{f,\varepsilon}; \mathbb{Z})) \supseteq \deg(g) \cdot \mathbb{Z}. \tag{6.11}$$

The proof is similar to that of Theorem 6.3, we only provide intuition regarding the appearance of the degree.

Proof (*Sketch*) Recall Chap. 5, by construction, there is smooth Lyapunov function V for A [49]. As A is of codimension-1, then by the Jordan–Brouwer seperation theorem e.g., see [61, p. 107], the levelset $V^{-1}(c)$, for sufficiently small $c > 0$, consists out of two components, think of $\mathsf{A} = \mathbb{S}^1 \subset \mathbb{R}^2$. Let V_c denote the "*outer*" component of $V^{-1}(c)$ and let A_δ be a tubular neighbourhood of A, which exists [78, Theorem 6.24]. Due to asymptotic stability one can pick $\delta > 0$ such that for all $x \in A_\delta \setminus \mathsf{A}$ one has $(x, \mu(x)) \in \Omega_{f,\varepsilon}$, moreover, by V being proper, there is $c > 0$ such that $V_c \subseteq A_\delta$. Now one can essentially copy the commutative diagram from the proof of Theorem 6.3, using the same functions, that is

$$H_{n-1}(V_c; \mathbb{Z}) \xrightarrow{g_\star} H_{n-1}(\mathbb{R}^n \setminus \{0\}; \mathbb{Z}) \simeq \mathbb{Z}$$
$$\downarrow{v_\star} \quad\quad\quad \nearrow^{f_\star}$$
$$H_{n-1}(\Omega_{f,\varepsilon}; \mathbb{Z})$$

Therefore we have that $f_\star(H_{n-1}(\Omega_{f,\varepsilon}; \mathbb{Z})) \supseteq g_\star(H_{n-1}(V_c; \mathbb{Z})$. Then, as V_c is one of the two components of $V^{-1}(c)$, $H_{n-1}(V_c; \mathbb{Z}) = H_0(V_c; \mathbb{Z}) \simeq \mathbb{Z}$ by Poincaré duality. Better yet, V_c is a closed, oriented manifold itself, such that $\deg(g)$ via $g_\star$ is well-defined, e.g., recall the rationale of Lemma 4.1 and consider the following diagram or see [82] for the full explanation.

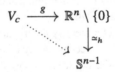

In contrast to Theorem 6.3, one does not assume that $\mu(a) = 0$ for all $a \in \mathsf{A}$. Indeed, assumptions of that form are not necessary, they rather (over)simplify the exposition.

Condition (6.11) is not particularly transparent, one rarely has access to $\deg(g)$. However, by the assumptions on A, one can appeal to the the *Gauss–Bonnet–Hopf (Dyck) theorem* [53, 88], stating that for a closed manifold $M^{n-1} \subset \mathbb{R}^n$ the *Gauss map* $\gamma : M^{n-1} \to \mathbb{S}^{n-1}$, defined by $\gamma(p)$ being the unit normal at $p \in M$, one has $\deg(\gamma) = \chi(N)$ for N the bounded component of $\mathbb{R}^n \setminus M^{n-1}$. In fact, this theorem relates directly to a manifestation of the Poincaré–Hopf theorem for manifolds *with* boundary and vector fields pointing outward cf. Sect. 8.2. Then, using the Gauss–Bonnet–Hopf theorem, one can relate the vector field g on the components of the levelset $V^{-1}(c)$ to inward- and outward pointing vector fields, normal to those manifolds. As such, this allows for linking $\deg(g)$ to the topology of A, e.g., due to a result by Hopf, when n is odd, $\deg(g) = \frac{1}{2}\chi(A)$ [82, Theorem 4].[14] Now one can show that the nonholonomic integrator cannot be stabilized to $\mathbb{S}^2$ either since $\chi(\mathbb{S}^2) = 2$ while the left-hand-side of (6.11) evaluates to 0 [82, Corollary 2]. A similar, but not identical, result can be shown for A being of codimension strictly larger than 1 [83].

The aforementioned results on submanifolds are of a local nature and generalize the condition due to Coron. Using retraction theory one can provide obstructions of a global nature and without explicit knowledge of vector fields.

Proposition 6.1 (Moulay and Bhat [92, Proposition 10]) *Let* M *be a smooth manifold and* A *a compact, embedded submanifold of* M. *Then* A *is a strong neighbourhood deformation retract of* M.

Ultimately, Theorem 6.12, as discussed in the next section, states that if a set $A \subseteq M$ is an attractor under a flow φ, then A is a weak deformation retract of $\mathcal{D}(\varphi, A)$. Therefore, in case A is an attractor, combining Proposition 6.1 with Lemma 2.2 and Theorem 6.12 leads to A being a strong deformation retract of $\mathcal{D}(\varphi, A)$. This implies in particular that A can only be a global attractor if it is a strong deformation retract of M. This result also appeared, with a different proof, in [130, Lemma 4]. As highlighted in [130], indeed, for A to be a global attractor of M, the spaces must be homotopy equivalent, e.g., stabilizing $\mathbb{S}^1$ in $\mathbb{S}^2$ fails as $\chi(\mathbb{S}^1) = 0$ while $\chi(\mathbb{S}^2) = 2$.

Returning to Example 6.6, consider two agents with $M - \mathbb{S}^1$ such that $(M)^? = \mathbb{T}^2$ and $C = \Delta_{\mathbb{S}^1} \simeq_t M$, although $\chi(\mathbb{S}^1) = 0$ and $\chi(\mathbb{T}^2) = 0$, C and $(M)^2$ are not homotopic as they have for instance different homology groups and different fundamental groups,[15] i.e., $\pi_1(\mathbb{S}^1) \simeq \mathbb{Z}$ and $\pi_1(\mathbb{T}^2) \simeq \mathbb{Z}^2$.

Proposition 6.1, however, says more, the deformation retract is of the *strong* type. For example, when M is an *absolute neighbourhood retract* (ANR),[16] then, A being a deformation retract of M is equivalent to A being a strong deformation retract of M [63, p. 199], see also [118, p. 31]. As before, we immediately observe some form of an odd-number limitation.

Corollary 6.2 (Odd-number limitation for attractors) *Let* M *be a smooth, closed manifold with* $\chi(M) \neq 0$, *then* M *does not admit any continuous vector field* X *such that any odd-dimensional, compact, embedded submanifold* A *of* M *is a global attractor under* X.

Using Proposition 6.1 and the mod 2 intersection theory from Section 3.4, Theorem 6.5 can be generalized beyond stabilization of a point. Again, this result is stated for simplicity using *vector* bundles. Generalizations to fiber bundles are possible.

---

[14] Informally, given a compact manifold M with $\partial M \neq \emptyset$, glue a copy of M to M itself along their boundaries and call the resulting manifold M', which is now closed. Then we have $\chi(M') = 2\chi(M) - \chi(\partial M)$. As such, when M is odd-dimensional, $\chi(M') = 0$ such that $\chi(M) = \frac{1}{2}\chi(\partial M)$. Formally, one employs a so-called *Mayer–Vietoris sequence* argument [118, Sect. 4.6].

[15] We did not touch upon *homotopy groups*, the fundamental group being the first homotopy group, but point the reader to [77, Chap. 7] for more information. Intuitively, the fundamental group of a topological space X measures how many different loops, up to homotopy, the space X admits.

[16] A Hausdorff space X is **normal** when for any two closed subsets $U, U' \subseteq X$ there are open neighbourhoods $W \supset U$, $W' \supset U'$ such that $W' \cap W = \emptyset$. Then, a space X is an **ANR** when for any embedding in a normal space $\iota : X \hookrightarrow Y$, $\iota(X)$ is a neighbourhood retract [13].

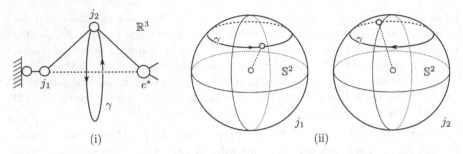

Fig. 6.5 Example 6.8, (i): robotic arm in $\mathbb{R}^3$ with γ the trajectory that keeps $e^\star$ fixed; (ii): the configuration space $\mathbb{S}^2 \times \mathbb{S}^2$ of the robot arm with the curves corresponding to γ, which are not independent as indicated by the marker on the curves

Theorem 6.10 (Obstruction to submanifold stabilization: bundles) *Let $\pi : \mathsf{M} \to \mathsf{B}$ be a vector bundle over some smooth, boundaryless, compact, connected manifold B and let $\mathsf{A} \subseteq Z_\pi(\mathsf{B})$ be a compact, embedded submanifold of M. If there is a continuous flow φ on M such that A is a global attractor under φ, then, $\mathsf{A} = Z_\pi(\mathsf{B})$.*

Proof The case where A is a point follows from Theorem 6.5 (Lemma 3.1 and Theorem 6.4). Regarding the general case, as demonstrated in Example 3.1, the zero section $Z_\pi(\mathsf{B})$ is a deformation retract of M. If A would be a global attractor, then, by Lemma 2.2, Proposition 6.1 and Theorem 6.12, A would be a deformation retract of M as well. In its turn, this implies by Lemma 2.1 that A is a deformation retract of $Z_\pi(\mathsf{B})$, which, by Lemma 3.2, cannot be true when $\mathsf{A} \neq Z_\pi(\mathsf{B})$.

Theorem 6.10 implies in particular that if M^m is a smooth, boundaryless, compact, connected manifold. Then, given any compact, embedded submanifold $\mathsf{A}^a \hookrightarrow \mathsf{M}^m$ with $0 \leq a < m$, there is no complete vector field $X \in \mathfrak{X}^{r \geq 0}(\mathsf{M}^m)$ such that A^a is a global attractor. Regarding Section 1.3 (compact Lie groups), not only *global* stabilization of a point is obstructed, but effectively of any non-trivial submanifold, as Lie groups are boundaryless. Then, as was the motivation for [9], Theorem 6.10 is of use in the context of mechanical systems. Recalling Example 2, a periodic orbit of the pendulum, as seen as a curve in $T\mathbb{S}^1$ is clearly homotopic to $\mathbb{S}^1$.

Example 6.8 (*Kinematic robot control*) Consider a two-link robotic arm with spherical joints. The goal is to globally stabilize the end-effector position at $e^\star \in \mathbb{R}^3$, see Fig. 6.5(i). Here we assume to work with a dynamical control system on $\mathsf{M} = T(\mathbb{S}^2 \times \mathbb{S}^2)$, i.e., a second-order system. As shown in Fig. 6.5(i), the configuration of the arm is not uniquely defined by $e^\star$, instead, one can freely move the elbow joint over the curve γ without changing the position of $e^\star$. One might want to exclude this ambiguity and render the dynamical system stationary on γ, while still globally stabilizing $e^\star$. As the motion of the two joints is not independent, the curve γ represents a 1-dimensional set in the configuration space $\mathbb{S}^2 \times \mathbb{S}^2$, e.g., see Fig. 6.5(ii). Regardless, one must be able to globally continuously stabilize a curve homeomorphic to $\mathbb{S}^1$ in the zero section of $\pi_p : T\mathbb{S}^2 \to \mathbb{S}^2$, for an individual joint, say j_1. This is prohibited by Theorem 6.10.

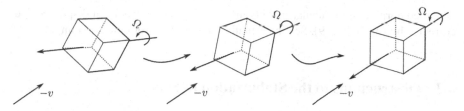

Fig. 6.6 Example 6.9: controlling a satellite (free rigid body) towards pointing along v, with angular velocity Ω, that is, the orientation is not fully fixed

Example 6.9 (*Satellite control*) Consider a free rigid body model of a satellite on $M^n = T\mathsf{SO}(3, \mathbb{R}) \simeq \mathsf{SO}(3, \mathbb{R}) \times \mathbb{R}^3$, where the last identification follows as Lie groups are parallelizable. Let $(R, \Omega) \in \mathsf{SO}(3, \mathbb{R}) \times \mathbb{R}^3$ denote the orientation and angular velocity, respectively, with $\Omega^\wedge \in \mathfrak{so}(3, \mathbb{R})$ cf. Example 5.2, then, for $\times$ the *cross-product*, an input-affine control system is given by

$$
\Sigma_{n,m}^{\mathrm{aff}} : \begin{cases} \dot{R}(t) = R(t)\Omega(t)^\wedge \\ I\dot{\Omega}(t) = I\Omega(t) \times \Omega(t) + \sum_{i=1}^{m} g_i u_i \end{cases}
\tag{6.12}
$$

for I some inertia tensor, $g_i \in \mathbb{R}^3$ and $u_i \in \mathbb{R}$ available inputs for $i = 1, 2, \ldots, m$, see [25], [11, p. 37]. Fix any $v \in \mathsf{S}^2 \subset \mathbb{R}^3$, we want to asymptotically stabilize the satellite, pointing along v, see Fig. 6.6. However, we do not specify a fixed rotation along this axis, only a constant velocity, that is, given the map $h : \mathsf{SO}(3, \mathbb{R}) \to \mathbb{R}^3$, $h : R \mapsto (R - I_n)v$ we want to render $\mathsf{A} = \{(R, v) \in \mathsf{M} : h(R) = 0\}$ globally asymptotically stable. Now, the set A is a 1-dimensional compact, embedded, submanifold of M as the set $\{R : h(R) = 0\}$ is a *stabilizer subgroup* of $\mathsf{SO}(3, \mathbb{R})$ with respect to v [45, p. 94], in particular, $\{R : h(R) = 0\}$ is isomorphic to $\mathsf{SO}(2, \mathbb{R}) \simeq \mathsf{S}^1$. As in [25], see that we do not demand the closed-loop vector field to vanish on A. Such a task might be *locally* feasible [89], however, for any m, continuous *global* asymptotic stabilization is obstructed by Theorem 6.10 (possibly, after a shift $\widetilde{\Omega} = \Omega - v$).

Similar topological obstructions occur for example in surgical robotics [125].

In [86], Mayhew and Teel extend Theorem 6.5 to the context of submanifold stabilization under set-valued maps. In particular, it is shown that for differential inclusions, i.e., vector fields of the form $\dot{x} \in X(x)$, $X : \mathsf{M} \rightrightarrows T\mathsf{M}$, satisfying the so-called "*basic conditions*" [86, Definition 5] the answer to Question (ii) is effectively the same as in the smooth case. The reason being that if a submanifold $\mathsf{A} \subseteq \mathsf{M}$ is an attractor under such a—possibly discontinuous—vector field, then there must exist a smooth, complete vector field X', defined on the domain of attraction of A, such that A is an attractor under X' [86, Theorem 14]. Although this framework captures some discontinuities, their conditions, however, do not capture for example Fig. 1.1(iii).

We end with a remark on compactness. In [127, Theorem 3.4] it is claimed that the domain of attraction, for *any* submanifold $\mathsf{A} \subseteq \mathsf{M}$, compact or non-compact, is homeomorphic to its tubular neighbourhood. This claim has been disproven and corrected by Lin et al. [79]. See in particular [79, Sect. 4] for counterexamples.

6.3 Obstructions to the Stabilization of Sets

Regarding Question (i), in 1993, Günther and Segal showed that a finite-dimensional compact set A can be an attractor of a continuous flow on a manifold if and only if A has the *shape*[17] of a polyhedron [55, Corollary 4]. Although the *Warsaw circle* is not homeomorphic to $\mathbb{S}^1$ it does have the shape of $\mathbb{S}^1$, see [58, Example 3.3] for an example by Hastings, rendering the Warsaw cirlce an attractor. Regarding realizable compact attractors, see also the article by Ortega and Sánchez–Gabites [97] and references therein. To add regarding Question (iii), given a compact attractor $A \subseteq M$ with domain of attraction $\mathcal{D}$, already the boundary of $\mathcal{D}$ can be arbitrarily complicated, cf. [110].

Hence, we pass to Question (ii). As mentioned before and indicated in [11, 82], if stabilization of a point is prohibited, stabilization of a set might be the next thing to consider. Concurrently, stabilization of a point might be to simplistic. In contrast to the previous section we will not assume this set to have a manifold structure, see that a variety of results in that section exploited this structure by appealing to the existence of a tubular neighbourhood.

Kvalheim and Koditschek recently generalized Brockett's condition to stabilization of any compact subset $A \subseteq M$ with $\chi(A) \neq 0$ [76]. To make sure $\chi(A)$ is well-defined, the authors appeal to *Čech–Alexander–Spanier* cohomology theory [76, Sect. 2], which is outside the scope of this exposition, but when discussing their result we assume to work with this homology theory.

Theorem 6.11 (Kvalheim–Koditschek condition [76, Theorem 3.2])
Let $\Sigma = (M, \mathcal{U}, F)$ be a control system and let $A \subseteq M$ be a compact attractor. Assume that $\chi(A) \neq 0$, then, for any neighbourhood $W \subseteq M$ of A there exists a neighbourhood $V \subseteq TM$ of the zero section of TM such that for any continuous vector field $X : W \to V$

$$F(\pi_u^{-1}(W)) \cap X(W) \neq \emptyset. \tag{6.13}$$

The motivation for Theorem 6.11 came from the development of repelling vector fields, i.e., to render some unsafe set $U \subset M$ repelling. Equivalently, one could render the safe set $S = M \setminus U$ attractive. After proving when $\chi(S)$ is well-defined, that result is summarized in [76, Theorem 3.6]. This result provides a variety of answers with respect to [130, Conjecture 2] stating that it is impossible for vehicles to smoothly converge to a desired configuration from every initial configuration in an environment scattered with obstacles, see [76, Example 2]. See also the work by Byrnes [24] for

---

[17] Shape theory is outside of the scope of this work, we refer the reader to the survey articles by Sánchez–Gabites [109] and Sanjurjo [112].

earlier generalizations of Brockett's condition with respect to the global stabilization of $A \subseteq \mathbb{R}^n$.

Now, recall Example 6.1. Mansouri showed that the cylinder can be rendered an attractor [82, p. 1]. Then, using Theorem 6.11 one can show that the topology of the cylinder is crucial here as no set with non-zero Euler characteristic can be stabilized. Interestingly, when A is a point, Theorem 6.11 can be shown to be strictly stronger than Brockett's condition, while the condition turns out to be weaker than Coron's condition [76, Sect. 6]. This trade-off is however a recurring one, conditions based on homology theory are frequently stronger, but also significantly harder to check.

Recall the definition of a dynamical system $(M, \mathbb{R}, \varphi)$ in the sense of Sect. 5.1. Then, using the retraction theory as set forth in Chap. 2, we can provide a generic result, a generalization of the work by Sontag [116, Theorem 21] (Theorem 6.4) and Bhatia and Szegö [10, Lemma V 3.2], due to Moulay and Bhat.

Theorem 6.12 (Moulay–Bhat condition [92, Theorem 5]) *Let $(M, \mathbb{R}, \varphi)$ be a dynamical system over a topological manifold M. Let $A \subseteq M$ be a compact attractor, with domain of attraction $\mathcal{D}(\varphi, A)$. Then A is a weak deformation retract of $\mathcal{D}(\varphi, A)$.*

Proof (*Sketch*) As a topological manifold is in particular a locally compact Hausdorff space, Theorem 5.1 applies. Now let $U_c = \{x \in \mathcal{D}(\varphi, A) : V(x) \leq c\}$. Note that U_c is positively invariant, i.e., $\psi^t(U_c) \subseteq U_c$ for all $t \geq 0$ and $\cap_{c>0} U_c = A$. Hence, for each open neighbourhood $W \subseteq M$ of A there is a $c > 0$ such that $U_c \subseteq W$. Now, define $T_c(x) = \inf\{t \geq 0 : \varphi^t(x) \in U_c\}$, which can be shown to be continuous. Then, pick any W with $c > 0$ such that $U_c \subset W$ and consider the map $H : I \times \mathcal{D}(\varphi, A) \to \mathcal{D}(\varphi, A)$ defined by $H(t, x) = \varphi(tT_c(x), x)$, which is continuous and satisfies $H(0, x) = x$ for all $x \in \mathcal{D}(\varphi, x)$, $H(1, x) \in U_c$ for all $x \in \mathcal{D}(\varphi, A)$ and $H(t, x) = x$ for all $x \in U_c$. Hence, H parametrizes a strong deformation retract from $\mathcal{D}(\varphi, A)$ onto U_c. As W was an arbitrary open neighbourhood of A, this shows that A is weak deformation retract of $\mathcal{D}(\varphi, A)$. See [92] for the details.

In combination with Lemma 2.2 one recovers for example from Theorem 6.12 that if $A \subseteq M$ is a compact attractor of a dynamical system $(M, \mathbb{R}, \varphi)$, then, if A is a strong neighbourhood retract of M, A is a strong deformation retract of its domain of attraction $\mathcal{D}(\varphi, A)$ [92, Corollary 7]. We emphasize that Theorem 6.12 remains true for M being a locally compact Hausdorff space.

Example 6.10 (*The rotation group as a potential attractor*) For any $A \in \mathsf{GL}^+(n, \mathbb{R})$ let $A = UP$ be its *Polar decomposition*, for some $U \in \mathsf{SO}(n, \mathbb{R})$ and symmetric positive definite $P \in S_{>0}^n$. Then consider the map $H(t, A) = U(tI_n + (1 - t)P)$ for $t \in [0, 1]$. See that $H(0, A) = A$ for all $A \in \mathsf{GL}^+(n, \mathbb{R})$, $H(1, A) = U$ for $A = UP$ and $H(t, U) = U$ for all $U \in \mathsf{SO}(n, \mathbb{R})$ and all t. Hence, as $A \mapsto U$ is continuous, $\mathsf{SO}(n, \mathbb{R})$ is a strong deformation retract of $\mathsf{GL}^+(n, \mathbb{R})$.[18] Now for some $B \in \mathsf{SL}(n, \mathbb{R})$, let $B = QR$ be its *QR decomposition* for $Q \in \mathsf{SO}(n, \mathbb{R})$ and $R \in \mathbb{R}^{n \times n}$ being upper-triangular with strictly positive elements on its diagonal. Now

---

[18] See that $U = A(\sqrt{A^\top A})^{-1}$. The reader is invited to visualize a slice of $\mathsf{SO}(2, \mathbb{R}) \hookrightarrow \mathsf{GL}^+(2, \mathbb{R})$.

define the map $H(t, B) = QZ(t)$ with $t \in [0, 1]$ for $Z(t)$ such that $z_{ii}(t) = r_{ii}^t$ and $z_{ij}(t) = tr_{ij}$ for $i \neq j$. Then we see that $H(0, B) = Q$ for $B = QR$, $H(1, B) = B$ and $H(t, Q) = Q$ for all $Q \in SO(n, \mathbb{R})$. Again, one can show that this decomposition is continuous such that $SO(n, \mathbb{R})$ is also a strong deformation retract of $SL(n, \mathbb{R})$. Note that for any $n > 1$, $SO(n, \mathbb{R})$ cannot be a global attractor of a flow on $\mathbb{R}^{n \times n} \simeq \mathbb{R}^{n^2}$ by Lemma 3.1. Similarly, although we can construct a trivial embedding $\iota : SO(n, \mathbb{R}) \hookrightarrow SO(n + 1, \mathbb{R})$, by Lemma 3.2 $SO(n, \mathbb{R})$ cannot be a global attractor in such an ambient space either. For further comments, see [21].

Remark 6.3 (*On a proof of Theorem 6.4*) Consider now A being a point, denoted p^*, and assume that M is locally contractible, as any topological manifold. Clearly, p^* is a strong neighbourhood retract of M. Hence, Theorem 6.4 follows as a corollary to Lemma 2.2 and Theorem 6.12. When $M = \mathbb{R}^n$, Theorem 6.12 can be strengthened to A being a *strong* deformation retract of $\mathcal{D}(\varphi, A)$ [7].

Remark 6.4 (*Stabilization of A with $\chi(A) = 0$*) Whereas works like [76, 82, 83] can in general not address the stabilization of compact sets $A \subseteq M$ with $\chi(A) = 0$, the retraction-based results provide some necessary conditions, see Theorem 6.12 and Example 6.10. A necessary condition for global stabilization of A of the form $\chi(A) = \chi(M)$ is clearly weaker than the deformation retract formulation, e.g., the preservation of the Euler characteristic is necessary for homotopy equivalence, but not sufficient. For instance, compare the homology groups of $\mathbb{S}^1 \hookrightarrow \mathbb{T}^2$ to that of the ambient space.

Remark 6.5 (*More on robustness*) The retraction-based results are robust in the sense that they are true for *any* continuous control system. However, consider stabilizing the unit disk $\mathbb{D}_1^n$ and the punctured unit disk $\mathbb{D}_1^n \setminus \{0\}$ in $\mathbb{R}^n$. Again, arbitrarily small perturbations can potentially invalidate a necessary condition for continuous stabilization. This is important to take into account with numerical methods in mind.

Coming back to where we started, Kvalheim recently generalized the *homological* results due to Coron and Mansouri via appealing to *homotopical* arguments in the spirit of those by Bobylev, Krasnosel'skiĭ and Zabreĭko indeed, cf. Example 3.2, Example 3.4 and [75]. Now, we recall Assumption 5.1 and state the first result.

Theorem 6.13 (Kvalheim's homotopy [75, Theorem 1]) *Let M be a smooth manifold and let $X, Y \in \mathfrak{X}^{r \geq 0}(M)$ be such that the compact set $A \subseteq M$ is asymptotically stable under both. Then, there is an open neighbourhood U of A such that $X|_{U \setminus A}$ and $Y|_{U \setminus A}$ are homotopic through non-vanishing vector fields.*

As in Example 3.4, using the flows corresponding to X and Y, locally, a homotopy is constructed to prove Theorem 6.13. Note that the tools from Chapter 4 now imply that for U as in Theorem 6.13 and any $W \subset U \setminus A$ we have that the following induced homomorphisms agree $(X|_W)_*, (Y|_W)_* : H_{(\cdot)}(W) \to H_{(\cdot)}(TW \setminus \{0\})$. This observation is exploited below.

The aforementioned results can now be unified as follows. One constructs a "*canonical*" vector field Y on an open neighbourhood U of A such that A is locally asymptotically stable. For this canonical vector field Y one computes some homotopy invariant of interest and by Theorem 6.13 this extends to any vector field locally asymptotically stabilizing A. The importance of having the vector fields to be non-vanishing on $U \setminus A$ is displayed in for instance Example 3.4 and the proof (sketch) of Theorem 6.3, without this requirement we are not capturing meaningful information, e.g., any two continuous vector fields on $\mathbb{R}^n$ are straight-line homotopic.

In the setting of an equilibrium point the canonical vector field (with the origin being asymptotically stable) is locally given by $\dot{x} = -x$. Indeed, from there one computes the corresponding index cf. Example 3.4. More general, for an embedded submanifold A in $\mathbb{R}^n$, the canonical stabilizing vector field is the negated normal vector field. Indeed, Mansouri exploits this and the existence of a tubular neighbourhood in [82, Theorem 4] to relate $\deg(g)$ from Theorem 6.9 to the underlying topology of A. In the general case, one cannot appeal to tubular neighbourhoods and the canonical vector field Y is less obvious to select. One usually passes to the dynamical system generated by the negative gradient flow of a Lyapunov function.

It is important to stress that all of these results provide *necessary*, but by no means *sufficient*, conditions, e.g., consider vector fields of the form $X_1 = X$ and $X_2 = -X$.

The same is true for the final Theorem of this chapter. Nevertheless, recall that $Y_\star$ and $F_\star$ denote induced homomorphisms and recall the notion of a control system Σ as given by Definition 5.5, then, one can use Theorem 6.13 to derive rather generic homological necessary conditions.

Theorem 6.14 (Kvalheim's condition [75, Theorem 2]) *Let* M *be a smooth manifold and let* $\Sigma = (M, \mathcal{U}, F)$ *be a continuous control system. Assume there is a* $Y \in \mathfrak{X}^{r \geq 0}(M)$ *such that the compact subset A of M is asymptotically stable under Y. Moreover, assume there is a feedback μ that asymptotically stabilizes A. Then, for a sufficiently small open neighbourhood U of A one has*

$$Y_\star H_{(\cdot)}(U \setminus A) \subseteq F_\star H_{(\cdot)}(\pi_u^{-1}(U \setminus A) \setminus F^{-1}(0)) \subseteq H_{(\cdot)}(T(U \setminus A) \setminus \{0\}). \quad (6.14)$$

In the proof of Theorem 6.14, one exploits the decomposition as $(F \circ \mu)_\star = F_\star \circ \mu_\star$. As pointed out in [75, Remark 1], the results due to Coron and Mansouri follow indeed from (6.14). Again, one can find control systems that do satisfy the conditions of Theorem 6.14, yet, continuous asymptotic stabilization is impossible. The reader is invited to find non-trivial examples. See also that Theorem 6.14 does *not* assume $\chi(A) \neq 0$. In conclusion,

(i) we see that retraction theory allows for the construction of necessary conditions independent of the precise continuous control Σ system, cf. Theorem 6.12;
(ii) we also see that the methodology as put forth in the monograph by Krasnosel'skiĭ and Zabreĭko allows for generalizations far beyond characterizing the continuous stabilization of equilibrium points in $\mathbb{R}^n$, cf. Theorem 6.14.

6.4 Other Obstructions

Besides the aforementioned topological obstructions, a topological viewpoint can be seen to be fruitful in other modern branches of system identification and control theory [16, 27, 66, 68, 81]. In particular, in [20] a topological obstruction to the *reach control problem* is presented. Obstructions to simultaneous stabilization (robust control), are considered in [41]. Extensions of Brockett's condition in the context of exponential stabilization are discussed in [56]. With respect to adaptive linear control, topological obstructions to self-tuning are presented by Polderman [100] and van Schuppen [122]. In the context of hybrid systems, Ames and Sastry present topological obstructions to zeno behaviour in [4]. In [84, 85], Mansouri presents topological obstruction to the existence of distributed controllers, that is, controllers where each input variable can only depend on a subset of the state variables.

Necessary and sufficient conditions for global, smooth, feedback linearization of a smooth, input-affine system (5.8) with $m = 1$ are presented in [39], for example, M must be simply connected, ruling out $SO(n, \mathbb{R})$ for $n \geq 2$.

Topological obstructions also appear in the context of *motion planning algorithms.* In line with Theorem 6.4, a globally defined continuous motion planning algorithm exists only if the underlying configuration space is contractible. See [47, 48], for more work by Farber on topological obstructions to motion planning algorithms and [52] for early remarks by Gottlieb. Somewhat related, see [80] for obstructions to certain tracking problems. In particular, see [80, Example 4.1] which considers Brockett's nonholonomic integrator (6.1).

There is also a line of work by, amongst others, Byrnes, Delchamps and Hazewinkel on the geometry and topology of linear systems, providing for further obstructions to for example global system identification, e.g., see [23, 26, 43].

See [29, Sect. 18.5] or [105, pp. 35–36] for related phenomena in the calculus of variations, the so-called "*Lavrentiev gap*".

In the context of physics, in particular particle physics, topological curiosities manifest themselves mathematically for example via Poincaré's lemma, e.g., the Ehrenberg–Siday–Aharonov–Bohm effect, and by means of the Atiyah–Singer index theorem, e.g., to understand spectral flows [44]. Here, the topological *obstruction* oftentimes relates to not being able to apply Stokes' theorem, e.g., a differential form fails to be exact. Earlier, topological obstructions were studied in the context of Hamiltonian mechanics [46].

Topological obstructions have also been reported in chemistry, although more related to data analysis [57]. For infinite-dimensional problems, topological obstructions frequently pertain to reachable sets being empty due to the fact that a compact set in an infinite-dimensional Banach space has empty interior, e.g., see [15]. Topological obstructions in the context of neural networks are alluded to in [128].

The topological viewpoint also provided to be useful early on in the context of Bellman equations [121], e.g., see the initial work by Petrov on regularity in the context of time-optimal processes [99].

References

1. Absil P-A, Mahony R, Sepulchre R (2009) Optimization algorithms on matrix manifolds. Princeton University Press, Princeton
2. Aeyels D (1985) Stabilization of a class of nonlinear systems by a smooth feedback control. Syst Control Lett 5(5):289–294
3. Amann A, Hooton EW (1999) An odd-number limitation of extended time-delayed feedback control in autonomous systems. Philos TR Soc A 371:2013
4. Ames AD, Sastry S (2005) A homology theory for hybrid systems: hybrid homology. International workshop on hybrid systems: computation and control. Springer, Berlin, pp 86–102
5. Arnold VI (1988) Mathematical methods of classical mechanics. Springer
6. Aydo A, Mcquade S, Duteil N (2017) Opinion dynamics on a general compact riemannian manifold. Netw. Heterog Media 12(3):489–523
7. Bernuau E, Moulay E, Coirault P, Hui Q (2019) Topological properties for compact stable attractors in $\mathbb{R}^n$. In: SIAM conference control applications, pp 15–21
8. Bernuau E, Perruquetti W, Moulay E (2013) Retraction obstruction to time-varying stabilization. Automatica 49(6):1941–1943
9. Bhat SP, Bernstein DS (2000) A topological obstruction to continuous global stabilization of rotational motion and the unwinding phenomenon. Syst Control Lett 39:63–70
10. Bhatia NP, Szegö GP (1970) Stability theory of dynamical systems. Springer, Berlin
11. Bloch A (2015) Nonholonomic mechanics and control. Springer, New York
12. Bloch A, Reyhanoglu M, McClamroch N (1992) Control and stabilization of nonholonomic dynamic systems. IEEE T Automat Conu 37(11):1746–1757
13. Borsuk K (1967) Theory of retracts. Państwowe Wydawn, Naukowe, Warszawa
14. Boumal N (2023) An Introduction to optimization on smooth manifolds. Cambridge University Press, Cambridge
15. Boussaid N, Caponigro M, Chambrion T (2019) On the ball-marsden-slemrod obstruction for bilinear control systems. In: Proceeding of IEEE conference on decision and control, pp 4971–4976
16. Bramburger JJ, Brunton SL, Kutz JN (2021) Deep learning of conjugate mappings. Phys D: Nonlinear Phenom 427:133008
17. Brockett RW (1982) Control theory and singular Riemannian geometry. New directions in applied mathematics. Springer, New York, pp 11–27
18. Brockett RW (1983) Asymptotic stability and feedback stabilization. Differential geometric control theory. Birkhäuser, Boston, pp 181–191
19. Brockett RW, Dai L (1993) Non-holonomic Kinematics and the role of elliptic functions in constructive controllability. Springer, Boston, pp 1–21
20. Broucke ME, Ornik M, Mansouri A-R (2017) A topological obstruction in a control problem. Syst Control Lett 108:71–79
21. Browne ML (1955) A note on the classical groups. Am Math Mon 62(6):424–427
22. Bullo F, Lewis AD (2004) Geometric control of mechanical systems. Springer, New York
23. Byrnes CI (1983) Geometric aspects of the convergence analysis of identification algorithms. Nonlinear stochastic problems. Springer, Dordrecht, pp 163–186
24. Byrnes CI (2008) On Brockett's necessary condition for stabilizability and the topology of Liapunov functions on $\mathbb{R}^n$. Commun Inf Syst 8(4):333–352
25. Byrnes CI, Isidori A (1991) On the attitude stabilization of rigid spacecraft. Automatica 27(1):87–95
26. Byrnes CI, Martin CF (1980) Geometrical methods for the theory of linear systems. D. Reidel Publishing Company, Dordrecht
27. Caines PE, Levanony D (2019) Stochastic ε-optimal linear quadratic adaptation: an alternating controls policy. SIAM J Contr Optim 57(2):1094–1126
28. Ceragioli F (2002) Some remarks on stabilization by means of discontinuous feedbacks. Syst Control Lett 45(4):271–281

29. Cesari L (2012) Optimization-theory and applications: problems with ordinary differential equations. Springer Science and Business Media, New York

30. Christensen F (2017) A necessary and sufficient condition for a unique maximum with an application to potential games. Econ Lett 161:120–123

31. Christopherson BA, Mordukhovich BS, Jafari F (2022) Continuous feedback stabilization of nonlinear control systems by composition operators. ESAIM: Contr Optim Ca 28:1–22

32. Cieliebak K, Eliashberg Y (2012) From Stein to Weinstein and back: symplectic geometry of affine complex manifolds. American Mathematical Society, Providence

33. Coron J-M (1990) A necessary condition for feedback stabilization. Syst Control Lett 14(3):227–232

34. Coron J-M (2007) Control and nonlinearity. American Mathematical Society, Providence

35. Coron J-M, Praly L (1991) Adding an integrator for the stabilization problem. Syst Control Lett 17(2):89–104

36. Coron J-M, Praly L, Teel A (1995) Feedback stabilization of nonlinear systems: sufficient conditions and Lyapunov and input-output techniques. Springer, London, pp 293–348

37. Coron J-M, Rosier L (1994) A relation between continuous time-varying and discontinuous feedback stabilization. J Math Syst Est Control 4:67–84

38. Dayawansa W (1993) Recent advances in the stabilization problem for low dimensional systems. Nonlinear control systems design 1992. Pergamon, Oxford, pp 1–8

39. Dayawansa W, Boothby W, Elliott D (1985) Global state and feedback equivalence of nonlinear systems. Syst Control Lett 6(4):229–234

40. Dayawansa W, Martin C, Knowles G (1990) Asymptotic stabilization of a class of smooth two-dimensional systems. SIAM J Contr Optim 28(6):1321–1349

41. Dayawansa WP, Qi B-M (1999) Topological obstructions to simultaneous stabilization. Proc IEEE Conf Decis Control 2:1196–1201

42. de Wolff B, Schneider I (2021) Geometric invariance of determining and resonating centers: odd-and any-number limitations of Pyragas control. Chaos 31(6):063125

43. Delchamps DF (1985) Global structure of families of multivariable linear systems with an application to identification. Math Syst Theory 18(1):329–380

44. Delplace P (2022) Berry-Chern monopoles and spectral flows. SciPost Phys Lect Notes 39

45. Duistermaat J, Kolk J (1999) Lie groups. Springer, Berlin

46. Duistermaat JJ (1980) On global action-angle coordinates. Comm Pure Appl Math 33(6):687–706

47. Farber M (2003) Topological complexity of motion planning. Discrete Comput Geom 29(2):211–221

48. Farber M (2004) Instabilities of robot motion. Topol Appl 140(2–3):245–266

49. Fathi A, Pageault P (2019) Smoothing Lyapunov functions. T Am Math Soc 371(3):1677–1700

50. Garay B (1991) Strong cellularity and global asymptotic stability. Fund Math 2(138):147–154

51. Goeleven D, Brogliato B (2005) Necessary conditions of asymptotic stability for unilateral dynamical systems. Nonlinear Anal Theor 61(6):961–1004

52. Gottlieb D (1986) Robots and topology. In: Proceedings IEEE international conference on robotics and automation, vol 3, pp 1689–1691

53. Gottlieb DH (1996) All the way with Gauss-Bonnet and the sociology of mathematics. Am Math Mon 103(6):457–469

54. Guillemin V, Pollack A (2010) Differential topology. American Mathematical Society, Providence

55. Günther B, Segal J (1993) Every attractor of a flow on a manifold has the shape of a finite polyhedron. Proc Am Math Soc 119(1):321–329

56. Gupta R, Jafari F, Kipka R, Mordukhovich B (2018) Linear openness and feedback stabilization of nonlinear control systems. Discrete Cont Dyn-S 11(6):1103–1119

57. Hashemian B, Arroyo M (2015) Topological obstructions in the way of data-driven collective variables. J Chem Phys 142(4):044102

58. Hastings H (1979) A higher-dimensional Poincaré-Bendixson theorem. Glas Mat 14(34):263–268
59. Hatcher A (2002) Algebraic topology. Cambridge University Press, Cambridge
60. Helmke U, Moore JB (2012) Optimization and dynamical systems. Springer, London
61. Hirsch MW (1976) Differential topology. Springer, New York
62. Hooton EW, Amann A (2012) Analytical limitation for time-delayed feedback control in autonomous systems. Phys Rev Lett 109(15):154101
63. Hu ST (1965) Theory of retracts. Wayne State University Press, Detroit
64. Isidori A (1985) Nonlinear control systems: an introduction. Springer, Berlin
65. Jean F (2014) Control of nonholonomic systems: from sub-Riemannian geometry to motion planning. Springer, Cham
66. Jongenccl W, Kuhn D (2014) On topological equivalence in Linear Quadratic optimal control. In: Proceeding IEEE European Control Conference pp 2002–2007
67. Jongeneel W, Schwan R (2023) On continuation and convex Lyapunov functions. arXiv e-prints ArXiv: 2301.05932
68. Jongeneel W, Sutter T, Kuhn D (2022) Topological linear system identification via moderate deviations theory. IEEE Contr Syst Lett 6:307–312
69. Jurdjevic V, Quinn JP (1978) Controllability and stability. J Differ Equ 28(3):381–389
70. Kawski M (1989) Stabilization of nonlinear systems in the plane. Syst Control Lett 12(2):169–175
71. Koditschek DE (1988) Application of a new Lyapunov function to global adaptive attitude tracking. In: Proceeding IEEE conference on decision and control, pp 63–68
72. Koditschek DE, Rimon E (1990) Robot navigation functions on manifolds with boundary. Adv Appl Math 11(4):412–442
73. Krasnosel'skiĭ A, Zabreiko PP (1984) Geometrical methods of nonlinear analysis. Springer, Berlin
74. Kurzweil J (1963) On the inversion of Ljapunov's second theorem on stability of motion. AMS Transl Ser 2(24):19–77
75. Kvalheim MD (2023) Obstructions to asymptotic stabilization. SIAM J Control Optim 61(2):536–542
76. Kvalheim MD, Koditschek DE (2022) Necessary conditions for feedback stabilization and safety. J Geom Mech
77. Lee JM (2011) Introduction to topological manifolds. Springer, New York
78. Lee JM (2012) Introduction to smooth manifolds. Springer, New York
79. Lin B, Yao W, Cao M (2022) On Wilson's theorem about domains of attraction and tubular neighborhoods. Syst Control Lett 167:105322
80. Lizárraga DA (2004) Obstructions to the existence of universal stabilizers for smooth control systems. Math Control Signals Syst 16(4):255–277
81. Lopez R, Atzberger PJ (2020) Variational autoencoders for learning nonlinear dynamics of physical systems. ArXiv preprint arXiv: 2012:03448
82. Mansouri A-R (2007) Local asymptotic feedback stabilization to a submanifold: topological conditions. Syst Control Lett 56(7–8):525–528
83. Mansouri A-R (2010) Topological obstructions to submanifold stabilization. IEEE T Automat Contr 55(7):1701–1703
84. Mansouri A-R (2013) Topological obstructions to distributed feedback stabilization. In: Annual Allerton Conference on Communication, Control, and Computing, pp 1573–1575
85. Mansouri A-R (2015) Topological obstructions to distributed feedback stabilization to a sub-manifold. In: SIAM Conference on Control and Its Applications, pp 76–80
86. Mayhew CG, Teel AR (2011) On the topological structure of attraction basins for differential inclusions. Syst Control Lett 60(12):1045–1050
87. McLennan A (2018) Advanced fixed point theory for economics. Springer, Singapore
88. Milnor J (1965) Topology from the differentiable viewpoint. Princeton University Press, Princeton

89. Montenbruck JM, Allgöwer F (2016) Asymptotic stabilization of submanifolds embedded in riemannian manifolds. Automatica 74:349–359
90. Mortensen RE (1968) A globally stable linear attitude regulator. Int J Control 8(3):297–302
91. Moulay E (2011) Morse theory and Lyapunov stability on manifolds. J Math Sci 177(3):419–425
92. Moulay E, Bhat SP (2010) Topological properties of asymptotically stable sets. Nonlinear Anal-Theor 73(4):1093–1097
93. Murray RM, Li Z, Sastry SS (1994) A mathematical introduction to robotic manipulation. CRC Press, Boca Raton
94. Nijmeijer H, van der Schaft A (1990) Nonlinear dynamical control systems. Springer, New York
95. Orsi R, Praly L, Mareels I (2003) Necessary conditions for stability and attractivity of continuous systems. Int J Control 76(11):1070–1077
96. Ortega R (1996) A criterion for asymptotic stability based on topological degree. In: Proceedings of the first world congress of nonlinear analysts, pp 383–394
97. Ortega R, Sánchez-Gabites JJ (2015) A homotopical property of attractors. Topol Method Nonl An 46(2):1089–1106
98. Palis JJ, De Melo W (1982) Geometric theory of dynamical systems: an introduction. Springer, New York
99. Petrov N (1970) On the Bellman function for the time-optimal process problem. J Appl Math Mech 34(5):785–791
100. Polderman JW (1987) Adaptive control and identification: conflict or conflux. Dissertation, University of Groningen
101. Polpitiya AD, Dayawansa WP, Martin CF, Ghosh BK (2007) Geometry and control of human eye movements. IEEE T Automat Contr 52(2):170–180
102. Pyragas K (1992) Continuous control of chaos by self-controlling feedback. Phys Lett A 170(6):421–428
103. Qian C, Lin W (2001) A continuous feedback approach to global strong stabilization of nonlinear systems. IEEE T Automat Contr 46(7):1061–1079
104. Reineck JF (1991) Continuation to gradient flows. Duke Math J 64(2):261–269
105. Rindler F (2018) Calculus of variations. Springer, Cham
106. Robinson C (1995) Dynamical systems: stability, symbolic dynamics, and chaos. CRC Press, Boca Raton
107. Ryan EP (1994) On Brockett's condition for smooth stabilizability and its necessity in a context of nonsmooth feedback. SIAM J Contr Optim 32(6):1597–1604
108. Salehi S, Ryan E (1985) A non-linear feedback attitude regulator. Int J Control 41(1):281–287
109. Sánchez-Gabites J (2008) Dynamical systems and shapes. RACSAM REV R Acad A 102(1):127–159
110. Sánchez-Gabites J, Sanjurjo J (2007) On the topology of the boundary of a basin of attraction. Proc Am Math Soc 135(12):4087–4098
111. Sanfelice RG, Messina MJ, Tuna SE, Teel AR (2006) Robust hybrid controllers for continuous-time systems with applications to obstacle avoidance and regulation to disconnected set of points. In: Proceedings of IEEE American control conference, pp 3352–3357
112. Sanjurjo JMR (2008) Shape and Conley index of attractors and isolated invariant sets. Differential equations, chaos and variationally problems. Basel, Birkhäuser, pp 393–406
113. Sastry S (1999) Nonlinear systems. Springer, New York
114. Sepulchre R, Aeyels D (1996) Homogeneous Lyapunov functions and necessary conditions for stabilization. Math Control Sig Syst 9(1):34–58
115. Smale S (1967) Differentiable dynamical systems. B Am Math Soc 73(6):747–817
116. Sontag ED (1998) Mathematical control theory: deterministic finite dimensional systems. Springer, New York
117. Sontag ED (2009) Stability and feedback stabilization. Springer, New York, pp 8616–8630
118. Spanier EH (1981) Algebraic topology. McGraw-Hill, New York
119. Takens F (1971). A solution. Manifolds-Amsterdam 1970. Springer, Berlin, pp 231–231

120. Tang A, Wang J, Low SH, Chiang M (2007) Equilibrium of heterogeneous congestion control: existence and uniqueness. IEEE/ACM T Netw 15(4):824–837
121. Vakhrameev S (1995) Geometrical and topological methods in optimal control theory. J Math Sci 76(5):2555–2719
122. van Schuppen J (1994) Tuning of Gaussian stochastic control systems. IEEE T Automat Contr 39(11):2178–2190
123. Vendittelli M, Oriolo G, Jean F, Laumond J-P (2004) Nonhomogeneous nilpotent approximations for nonholonomic systems with singularities. IEEE T Automat Contr 49(2):261–266
124. Wan C-J, Bernstein DS (1995) Nonlinear feedback control with global stabilization. Dynam Control 5(4):321–346
125. Webster RJ III, Kim JS, Cowan NJ, Chirikjian GS, Okamura AM (2006) Nonholonomic modeling of needle steering. Int J Robot Res 25(5–6):509–525
126. Wilson FW (1969) Smoothing derivatives of functions and applications. T Am Math Soc 139:413–428
127. Wilson FW Jr (1967) The structure of the level surfaces of a Lyapunov function. J Differ Equ 3(3):323–329
128. Winterhalder R, Bellagente M, Nachman B (2021) Latent space refinement for deep generative models. ArXiv e-prints ArXiv: 2106:00792
129. Yamamoto S, Ushio T (2003) Odd number limitation in delayed feedback control. Chaos control, pp 71–87
130. Yao W, Lin B, Anderson B, Cao M (2022) Topological analysis of vector-field guided path following on manifolds. IEEE T Automat Contr, 1–16
131. Zabczyk J (1989) Some comments on stabilizability. Appl Math Opt 19(1):1–9

Chapter 7
Towards Accepting and Overcoming Topological Obstructions

Obstruction results are aided by the generality offered by the topological viewpoint. Constructive results, however, oftentimes require more structure of the problem to be available [12], e.g., in 1983, Artstein [5] and Sontag [52] introduced CLFs; using center manifold theory, Aeyels was one of the first to provide constructive arguments towards smooth stabilization of nonlinear systems [1]; feedback linearization emerged in the late 1980s, as can be found in the monographs [27, 40]; and in the late 1980s backstepping was initiated by Byrnes, Isidori, Kokotović, Tsinias, Saberi, Sontag, Sussmann and others [31]. See also the 1985 and the 2001 surveys on constructive nonlinear control by Kokotović et al. [29, 30].

Regardless of these constructive methods, the last chapter showcased a collection of fundamental topological obstructions to the asymptotic stabilization of subsets of manifolds by means of merely continuous feedback, let alone smooth feedback. Therefore, in this chapter we briefly review a variety of proposed methods to deal with this situation. See for example [20] for a survey from 1995 on handling an assortment of topological obstructions. Since then, the focus shifted from smooth feedback to several manifestations of discontinuous techniques, as highlighted below.

7.1 On Accepting the Obstruction

Motivated by the Poincaré–Hopf theorem, early remarks on *almost* global asymptotic stabilization can be found in [28]. This notion is intimately related to multistability, however, now a *single* point *is* asymptotically stable whereas the rest is not, recall Fig. 1.1(iv). This relates to Question (iii) from Chap. 6, i.e., how does the dynamical system behave outside of the domain of attraction of the attractor under consideration? For example, when stabilizing an isolated equilibrium point on a closed manifold M^n by means of continuous feedback, the material from Chap. 6 implies that the remaining vector field indices must add up to $\chi(M^n) - (-1)^n$.

© The Author(s) 2023
W. Jongeneel and E. Moulay, *Topological Obstructions to Stability and Stabilization*,
SpringerBriefs in Control, Automation and Robotics,
https://doi.org/10.1007/978-3-031-30133-9_7

Theoretically handling multistability is usually done via an alteration of classical Lyapunov theory [4, 23], e.g., by passing to the so-called "*dual density formulation*" as proposed in [44]. There, global requirements are relaxed to *almost* global requirements. By doing so, topological obstructions are surmounted at the cost of potentially introducing singularities. This is what we saw in Sect. 1.3. This approach is for example illustrated in [3], where the inverted pendulum is almost globally stabilized.

Although almost global stability of some equilibrium point $p^\star$ can be justified by the points *not* in the domain of attraction of $p^\star$ being of measure zero, this is only true in the idealized setting. For instance, if one needs to take uncertainties or perturbations into account, even if arbitrarily small, this set to avoid remains by no means of measure zero. Hence, this approach cannot be categorized as robust.

7.2 On Time-Varying Feedback

Recall from Example 6.4 that time-varying feedback does not in general allow for overcoming *global* topological obstructions. Nevertheless, elaborating on the work by Sontag and Sussmann [51], under controllability assumptions, the so-called "*return method*" as devised by Coron shows that time-varying feedback can overcome some *local* topological obstructions to the stabilization of equilibrium points [17, 18]. Prior to the work by Coron, Samson showed that the nonholonomic integrator (6.1) can be stabilized by time-varying feedback indeed [39, 48, p. 566]. A more general and explicit approach is described in [41] by Pomet. We follow Sepulchre, Wertz and Campion [50] in providing intuition regarding this matter.

Example 7.1 (*On periodic feedback* [50]) Consider the nonlinear dynamical control system

$$\begin{cases} \dot{x}_1 = u, \\ \dot{x}_2 = x_1, \\ \dot{x}_3 = x_1^3. \end{cases} \tag{7.1}$$

System (7.1) is small-time-locally-controllable at 0 as, after writing (7.1) in standard form $\dot{x} = f(x) + g(x)u$, one finds that the set

$$\{g(x), [f(x), g(x)], [f(x), [f(x), [f(x), g(x)]]]\}$$

evaluated at 0, spans $\mathbb{R}^3 \simeq T_0\mathbb{R}^3$ [55]. Nonetheless, (7.1) fails to satisfy Brockett's condition, cf. Theorem 6.1. In particular, see that if $\mu(x_1, x_2, x_3)$ is any continuous state feedback aimed at asymptotically stabilizing 0, then μ cannot vanish on $A_\epsilon = \{x \in \mathbb{R}^3 : x_1 = 0, 0 < \|x\| < \epsilon\}$, for some sufficiently small $\epsilon > 0$. Otherwise, an additional equilibrium point would be introduced. This implies that μ cannot change sign on the annulus A_ϵ. Evidently, a change in sign might be necessary to stabilize a neighbourhood around the origin, for example consider $x(0) = (0, x_2(0), 0)$ with either $x_2(0) > 0$ or $x_2(0) < 0$. Hence, time-varying periodic feedback seems a viable option indeed. This, to enforce a (persistent) change in sign. One can construct a similar story regarding Brockett's nonholonomic integrator (6.1). To provide a hand-waving comment, let the system start from $(0, 0, x_3(0))$, with $x_3(0) \neq 0$. Then, both u_1, u_2 must become nonzero, pushing both x_1, x_2 away from 0. To make sure x_1, x_2 go back to 0, both u_1, u_2 must flip sign, but this zero crossing (easily) induces an equilibrium away from 0.

For more historical context, see the early survey paper [39] or the remarks in [19].

Regarding Sect. 1.3, with the aforementioned in mind, an optimal control cost akin to (1.2) might not be ideal as time is penalized while (time-invariant) continuous globally asymptotically stabilizing feedback is obstructed by $\chi(\mathsf{G}) = 0$. Moreover, the almost globally asymptotically stabilizing feedback is not robust.

As many examples before, including Example 7.1, highlight, is that some notion of switching is instrumental in overcoming topological obstructions. The next section highlights one of the most influential solution frameworks, discontinuous- and in particular, *hybrid control*. This framework is also capable of overcoming one of the inherent drawbacks of stabilizing (periodic) time-varying controllers: they are not robust, often slow and thereby costly.

7.3   On Discontinuous Control

As mentioned before, allowing for discontinuities in dynamical control systems is not an immediate remedy for topological obstructions, they can prevail [13, 37, 47]. Yet, under controllability assumptions—fitting to discontinuous solution frameworks, discontinuous feedback frequently allows for stabilization [2, 16, 33, 34], e.g., to control the nonholonomic integrator (6.1), one can consider a sliding-mode controller [9]. Notably, the CLF generalization due to Rifford, allows for a principled approach to designing stabilizing discontinuous feedback laws [45, 46], e.g., an explicit feedback stabilizing the nonholonomic integrator (6.1) is presented in [34]. See also [11] for a *numerical* study of CLFs with regard to the nonholonomic integrator. It is imperative to remark that under controllability assumptions, one can show that discontinuous stabilization schemes exist that are robust against measurement noise [53]. For an extensive tutorial paper on discontinuous dynamical systems, see [21], in particular the section on different notions of solutions. The survey articles by Clarke, on the other hand, focus more on control theoretic aspects [14, 15].

7.3.1   Hybrid Control Exemplified

Motivated by the modelling of physical systems, e.g., relays, Witsenhausen was one of the early contributors to hybrid control theory [57]. In part due to the aforemen-

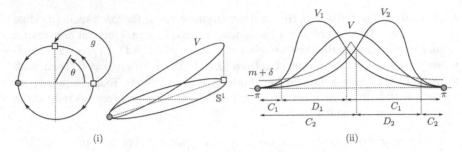

Fig. 7.1 For Example 7.2, (i) the jump map g and the Lyapunov function V and for Example 7.3, (ii) the family of Lyapunov functions, flow- and jump sets

tioned topological obstructions, however, hybrid control theory became an abstract theory of its own, e.g., see [25, 49, 56]. In this section we will only scratch the surface (Lyapunov-based switching) of what is possible using these techniques, e.g., in general one would discuss differential inclusions, hysteresis and so forth.

Example 7.2 (*Example 1 continued*) Recall Fig. 1.1(iii), we start by introducing how one could study discontinuous dynamical systems. Consider a **hybrid dynamical system** on the circle $\mathbb{S}^1 \subset \mathbb{R}^2$

$$\mathcal{H} : \begin{cases} \dot{x} = f(x), & x \in C \\ x^+ = g(x), & x \in D \end{cases} \tag{7.2}$$

for $C \subseteq \mathbb{S}^1$ the *flow set* and $D \subseteq \mathbb{S}^1$ the *jump set*, with g the corresponding *jump map*. This means that on the set C, the system behaves as the dynamical systems we encountered before, but on D, the state could possibly change in a way that one cannot describe using C^0 vector fields and flows. For instance, a hybrid system might describe a walking robot with the jump map handling the impact with the ground. See [25] for how to go about solutions of hybrid systems, or see [42, Sect. 24.2] for a succinct introduction. Now, parametrizing a hybrid dynamical system in polar coordinates on $\mathbb{S}^1$, consider $\dot{\theta} = \sin(\theta)$ under the jump map $g(\theta) = \varepsilon$ for some $\varepsilon \in (0, \pi)$ with the jump set being $D = \{0\}$. Then, by means of the Lyapunov function $V(\theta) = \cos(\theta) + 1$ stability of this hybrid system can be asserted since $\langle \partial_\theta V(\theta), \dot{\theta} \rangle = -\sin(\theta)^2$ while for all $\theta \in D$ one has $V(g(\theta)) < V(\theta)$. This is graphically summarized in Fig. 7.1(i). Again, we refer to [25, 42] for the technicalities of stability in the hybrid context.

The approach as taken in Example 7.2 lacks robustness and is too ad-hoc for practical purposes. In the context of stabilization of an equilibrium point $p^\star \in \mathsf{M}$, a successful hybrid methodology to overcome topological obstructions is to exploit *multiple* potential (Lyapunov) functions [10]. The so-called "*synergistic*" approach uses a family of potential functions $\mathcal{V} = \{V_i\}_{i \in \mathcal{I}}$, $V_i : \mathsf{M} \to \mathbb{R}_{\geq 0}$, with $p^\star$ being a critical point of all V_i. The remaining critical points can be different, but if $q^\star \neq p^\star$

is a critical point of V_j there must be a $k \neq j$ such that $V_k(q^*) < V_j(q^*)$. Now, if one switches appropriately between controllers induced by these potential functions, one cannot get trapped at the wrong critical point and hence p^* will be stabilized.

Example 7.3 (*Example 7.2 continued*) We will now follow [42, Example 24.5] and consider a hybrid *control* system on the embedded circle $\mathbb{S}^1 \subset \mathbb{R}^2$ and sketch more formally how the synergistic method works. Consider the control system $\dot{x} = u\Omega x$, where $u \in \mathbb{R}$ and $\Omega \in \mathsf{Sp}(2, \mathbb{R})$ such that $\Omega x \in T_x \mathbb{S}^1 = \{v \in \mathbb{R}^2 : \langle v, x \rangle = 0\}$. The goal is the same as before, to globally stabilize $\theta = \pi$, e.g., $x' = (-1, 0) \in \mathbb{R}^2$. We assume to have a smooth potential function $V : \mathbb{S}^1 \to \mathbb{R}_{\geq 0}$ with two critical points, x' at its global minimum and some other point $\bar{x}$ at its global maximum. Then, this potential function V is used to construct the family of potential functions $\mathscr{V} = \{V_1, V_2\}$, as discussed above, see [35] for the details. To be able to switch between these functions, the state space is augmented with the set $\{1, 2\}$. Now let $m(x) = \min_{q \in \{1,2\}} V_q(x)$, $C = \{(x, q) \in \mathbb{S}^1 \times \{1, 2\} : m(x) + \delta \geq V_q(x)\}$ and $D = \{(x, q) \in \mathbb{S}^1 \times \{1, 2\} : m(x) + \delta \leq V_q(x)\}$ for some $\delta > 0$, that is, we are in the jump set when the current selection of V_q is "*too large*". Now define the (set-valued) map $g_q : \mathbb{S}^1 \rightrightarrows \{1, 2\}$ by $g_q(x) = \{q \in \{1, 2\} : V_q(x) = m(x)\}$ and select the feedback controller, i.e., the input u, as

$$\mu(x, q) = -\langle \operatorname{grad} V_q(x), \Omega x \rangle. \tag{7.3}$$

Indeed, see that under (7.3), V_q is a Lyapunov function for x'_q on $\mathbb{S}^1 \setminus \{\bar{x}_q\}$ as $\dot{V}_q = -\langle \operatorname{grad} V_q(x), \Omega x \rangle^2 < 0$ for all $x \in \mathbb{S}^1 \setminus \{\bar{x}_q, x'_q\}$. Note, we exploit the embedding $\mathbb{S}^1 \hookrightarrow \mathbb{R}^2$. Summarizing, we have the closed-loop hybrid system

$$\mathcal{H} : \begin{cases} \dot{x} = \mu(x, q)\Omega x, \ \dot{q} = 0, & (x, q) \in C \\ x^+ = x, & q^+ = g_q(x), \ (x, q) \in D \end{cases} \tag{7.4}$$

For the technicalities, in particular how to define δ, we refer to [35], see also Fig. 7.1(ii). Also note that by construction the stabilization is robust, which should be contrasted to closed-loop systems as given in Example 7.2.

See for example [36, 38] for constructive results on $\mathsf{SO}(3, \mathbb{R})$, [42] for more on Example 7.3 and material concerning hybrid reinforcement learning and see [6, 54] for the hybrid approach in the context of optimization on manifolds.

A particular instance of a hybrid dynamical system is a **switched dynamical system** [32], [25, Sects. 1.4 and 2.4], that is, a dynamical system of the form $\dot{x} = f_\sigma(x)$ for some switching signal σ e.g., $\sigma : \operatorname{dom}(f) \to \{1, 2, \dots, N\}$, $N \in \mathbb{N}$. As alluded to in Figure 1.1, introducing a switch, that is, a discontinuity, can allow for global stabilization. Indeed, (7.4) is a manifestation of a switched system. See for example [32, pp. 87–88] for a switching-based solution to stabilization of the nonholonomic integrator (6.1). Slightly switching gears, a way to overcome topological obstructions is to construct a set of local controllers such that the union of their respective domains of attraction covers the space M. This does not necessarily result in global stabilization. The aim is rather to rule out instability by employing a suitable switching mechanism. Assuming that these controllers are intended to stabilize contractible sets, using the Lusternik–Schnirelmann category as will be introduced in Sect. 8.4, one can bound the number of required controllers from below. Clearly, on $\mathbb{S}^1$ one needs at least two of those controllers.

In contrast to almost global stabilization techniques, under controllability assumptions, for compact sets A, there always exists a hybrid controller that renders A locally asymptotically stable, in a robust sense [43]. The intuition being that measurable functions can be approximated by piecewise-continuous functions and converse Lyapunov results can be established for hybrid dynamical systems [24].

7.3.2 Topological Perplexity

As continuous feedback is frequently prohibited, one might be interested in obtaining a lower bound on the *"number"* of discontinuous actions required to render a set globally asymptotically stable. For the global stabilization of a point p on $\mathbb{S}^1$, this number is clearly 1 cf. Figure 1.1(iii). Recent work by Baryshnikov and Shapiro sets out to quantify this for generic spaces [7, 8]. The intuition is as follows, consider the desire to stabilize a point p on the torus $\mathbb{T}^2$ and recall its non-trivial singular homology groups

$$H_k(\mathbb{T}^2; \mathbb{Z}) \simeq \begin{cases} \mathbb{Z} & \text{if } k \in \{0, 2\} \\ \mathbb{Z} \oplus \mathbb{Z} & \text{if } k = 1 \end{cases}.$$

To globally stabilize a point, $\mathbb{T}^2$ must be deformed to a contractible space. Then, roughly speaking, as $H_1(\mathbb{R}^2; \mathbb{Z}) \simeq 0$, this can be achieved, for example, via two one-dimensional cuts. These cuts correspond to the discontinuities one needs to introduce to globally stabilize p on $\mathbb{T}^2$. Now if one desires to stabilize $\mathbb{S}^1 \hookrightarrow \mathbb{T}^2$, we need a single one-dimensional cut as $H_1(\mathbb{S}^1 \times \mathbb{R}; \mathbb{Z}) \simeq \mathbb{Z}$. Baryshnikov et al. proposes a method to quantify this approach merely based on topological information, so independent of metrics and coordinates. See also the algorithmic work [22].

In a similar vein, one can consider the work by Gottlieb and Samaranayake on indices of discontinuous vector fields [26]. The intuition being that Definition 1 considers a topological sphere around 0, in that sense there is no difference between Fig. 1.1(iii), (iv), i.e., one considers merely the boundary of a neighbourhood around the discontinuity.

References

1. Aeyels D (1985) Stabilization of a class of nonlinear systems by a smooth feedback control. Syst Control Lett 5(5):289–294

2. Ancona F, Bressan A (1999) Patchy vector fields and asymptotic stabilization. ESAIM Contr Optim Ca 4:445–471
3. Angeli D (2001) Almost global stabilization of the inverted pendulum via continuous state feedback. Automatica 37(7):1103–1108
4. Angeli D (2004) An almost global notion of input-to-state stability. IEEE T Automat Contr 49(6):866–874
5. Artstein Z (1983) Stabilization with relaxed controls. Nonlinear Anal Theor 7(11):1163–1173
6. Baradaran M, Poveda JI, Teel AR (2019) Global optimization on the sphere: a stochastic hybrid systems approach. IFAC 52(16):96–101
7. Baryshnikov Y (2021) Topological perplexity in feedback stabilization
8. Baryshnikov Y, Shapiro B (2014) How to run a centipede: a topological perspective. Geometric control theory and sub-riemannian geometry. Springer, Cham, pp 37–51
9. Bloch A, Drakunov S (1994) Stabilization of a nonholonomic system via sliding modes. In: Proceedings of IEEE conference on decision and control, vol 3, pp 2961–2963
10. Branicky MS (1998) Multiple Lyapunov functions and other analysis tools for switched and hybrid systems. IEEE T Automat Contr 43(4):475–482
11. Braun P, Grüne L, Kellett CM (2017) Feedback design using nonsmooth control Lyapunov functions: a numerical case study for the nonholonomic integrator. In: Proceedings of IEEE conference on decision and control, pp 4890–4895
12. Casti JL (1982) Recent developments and future perspectives in nonlinear system theory. SIAM Rev 24(3):301–331
13. Ceragioli F (2002) Some remarks on stabilization by means of discontinuous feedbacks. Syst Control Lett 45(4):271–281
14. Clarke F (2001) Nonsmooth analysis in control theory: a survey. Eur J Control 7(2–3):145–159
15. Clarke F (2004) Lyapunov functions and feedback in nonlinear control. Optimal control. Stabilization and nonsmooth analysis. Springer, Berlin, pp 267–282
16. Clarke FH, Ledyaev YS, Sontag ED, Subbotin AI (1997) Asymptotic controllability implies feedback stabilization. IEEE T Automat Contr 42(10):1394–1407
17. Coron J-M (1992) Global asymptotic stabilization for controllable systems without drift. Math Control Signals Syst 5(3):295–312
18. Coron J-M (1995) On the stabilization in finite time of locally controllable systems by means of continuous time-varying feedback law. SIAM J Contr Optim 33(3):804–833
19. Coron J-M (2007) Control and nonlinearity. American Mathematical Society, Providence
20. Coron J-M, Praly L, Teel A (1995) Feedback stabilization of nonlinear systems: sufficient conditions and Lyapunov and input-output techniques. Springer, London, pp 293–348
21. Cortes J (2008) Discontinuous dynamical systems. IEEE Contr Syst Mag 28(3):36–73
22. de Verdière ÉC (2012) Topological algorithms for graphs on surfaces. Habilitation, École normale supérieure
23. Efimov D (2012) Global Lyapunov analysis of multistable nonlinear systems. SIAM J Contr Optim 50(5):3132–3154
24. Goebel R, Hespanha J, Teel AR, Cai C, Sanfelice R (2004) Hybrid systems: generalized solutions and robust stability1. IFAC Proc 37(13):1–12
25. Goebel R, Sanfelice RG, Teel AR (2012) Hybrid dynamical systems. In: Hybrid dynamical systems. Princeton University Press, Princeton
26. Gottlieb DH, Samaranayake G (1995) The index of discontinuous vector fields. NYJ Math 1:130–148
27. Isidori A (1985) Nonlinear control systems: an introduction. Springer, Berlin
28. Koditschek DE (1988) Application of a new Lyapunov function to global adaptive attitude tracking. In: Proceedings of IEEE conference on decision and control, pp 63–68
29. Kokotović P, Arcak M (2001) Constructive nonlinear control: a historical perspective. Automatica 37(5):637–662
30. Kokotović PV (1985) Recent trends in feedback design: an overview. Automatica 21(3):225–236

31. Kokotović PV (1992) The joy of feedback: nonlinear and adaptive. IEEE Control Syst Mag 12(3):7–17
32. Liberzon D (2003) Switching in systems and control. Birkhäuser, Boston
33. Malisoff M, Krichman M, Sontag E (2006) Global stabilization for systems evolving on manifolds. J Dyn Control Syst 12(2):161–184
34. Malisoff M, Rifford L, Sontag E (2004) Global asymptotic controllability implies input-to-state stabilization. SIAM J Contr Optim 42(6):2221–2238
35. Mayhew CG, Teel AR (2010) Hybrid control of planar rotations. In: Proceedings of IEEE American control conference, pp 154–159
36. Mayhew CG, Teel AR (2011) Hybrid control of rigid-body attitude with synergistic potential functions. In: Proceedings of IEEE American control conference, pp 287–292
37. Mayhew CG, Teel AR (2011) On the topological structure of attraction basins for differential inclusions. Syst Control Lett 60(12):1045–1050
38. Mayhew CG, Teel AR (2011) Synergistic potential functions for hybrid control of rigid-body attitude. In: Proceedings of IEEE American control conference, pp 875–880
39. Morin P, Pomet J-B, Samson C (1998) Developments in time-varying feedback stabilization of nonlinear systems. IFAC Proc 31(17):565–572
40. Nijmeijer H, van der Schaft A (1990) Nonlinear dynamical control systems. Springer, New York
41. Pomet J-B (1992) Explicit design of time-varying stabilizing control laws for a class of controllable systems without drift. Syst Control Lett 18(2):147–158
42. Poveda JI, Teel AR (2021) A hybrid dynamical systems perspective on reinforcement learning for cyber-physical systems: vistas, open problems, and challenges. Handbook of reinforcement learning and control. Springer, New York, pp 727–762
43. Prieur C, Goebel R, Teel AR (2007) Hybrid feedback control and robust stabilization of nonlinear systems. IEEE T Automat Contr 52(11):2103–2117
44. Rantzer A (2001) A dual to Lyapunov's stability theorem. Syst Control Lett 42(3):161–168
45. Rifford L (2000) Existence of Lipschitz and semiconcave control-Lyapunov functions. SIAM J Contr Optim 39(4):1043–1064
46. Rifford L (2002) Semiconcave control-Lyapunov functions and stabilizing feedbacks. SIAM J Contr Optim 41(3):659–681
47. Ryan EP (1994) On Brockett's condition for smooth stabilizability and its necessity in a context of nonsmooth feedback. SIAM J Contr Optim 32(6):1597–1604
48. Samson C (1991) Velocity and torque feedback control of a nonholonomic cart. In: Advanced robot control, Berlin. Springer, pp 125–151
49. Sanfelice RG (2020) Hybrid feedback control. Princeton University Press, Princeton
50. Sepulchre R, Wertz V, Campion G (1992) Some remarks about periodic feedback stabilization. In: IFAC symposium on nonlinear control systems, pp 418–423
51. Sontag E, Sussmann H (1980) Remarks on continuous feedback. In: Proceedings of IEEE conference decision control, pp 916–921
52. Sontag ED (1983) A Lyapunov-like characterization of asymptotic controllability. SIAM J Control Optim 21(3):462–471
53. Sontag ED (1999) Clocks and insensitivity to small measurement errors. ESAIM Contr Optim Ca 4:537–557
54. Strizic T, Poveda JI, Teel AR (2017) Hybrid gradient descent for robust global optimization on the circle. In: Proceeding of IEEE conferences decision control, pp 2985–2990
55. Sussmann HJ (1987) A general theorem on local controllability. SIAM J Contr Optim 25(1):158–194
56. Van Der Schaft AJ, Schumacher JM (2000) An introduction to hybrid dynamical systems. Springer, London
57. Witsenhausen H (1966) A class of hybrid-state continuous-time dynamic systems. IEEE T Automat Contr 11(2):161–167

Chapter 8
Generalizations and Open Problems

8.1 Comments on Discrete-Time Systems and Periodic Orbits

We largely focused on vector fields, yet, in Chap. 3 one observes that a variety of results are shown via passing to the corresponding flow. In particular, recall the Lefschetz theory, i.e., Theorem 3.5. Akin to Example 3.4, one can compute the Lefschetz number for an asymptotically stable fixed point, e.g., recall (3.7). Consider the (time-one) map $x \mapsto Ax$ with $A \in \mathbb{R}^{n \times n}$ being Schur stable, then $\operatorname{sgn} \det(A - I_n) = (-1)^n$. However, instead of continuous-time dynamical systems of the form $(\mathsf{M}, \mathbb{R}, \varphi)$ one might be interested in *discrete-time* dynamical systems of the form $(\mathsf{M}, \mathbb{Z}, \psi)$, e.g., one works with discrete measurements of the state variables. Note, here the ***time-one map*** $\psi^1 : \mathsf{M} \to \mathsf{M}$ is enough to describe the dynamical system. Now, following [4, Sect. 1.11], we remark that such a discrete-time system corresponds directly to a continuous-time system. Define the *ceiling function* $c : p \in \mathsf{M} \mapsto 1$ and let $\mathsf{M}_1 = \{(p, s) \in \mathsf{M} \times [0, c(p)]\}/\sim$ for the equivalence relation $(p, c(p)) \sim (\psi^1(p), 0)$. Indeed, M_1 can be a Klein bottle (see Fig. 8.1i) when $\mathsf{M} = \mathbb{S}^1$. Then, one can define the semiflow $\phi^t : \mathsf{M}_1 \to \mathsf{M}_1$, said to be *a **suspension** of* ψ^1 under c, by $\phi^t(p, s) = (\psi^n(p), s')$ where the free variables n and s' need to satisfy $n + s' = t + s$ and $0 \leq s' \leq 1$. The manifold M_1 is also called the ***mapping torus*** of ψ^1. One should observe that although M_1 and M are not homotopic, by identifying a set $A \subset \mathsf{M}$ with a set $A' \subset \mathsf{M}_1$, one can study topological obstructions in discrete-time via passing to the continuous-time. We note, however, that practically, hybrid models might be preferred over purely discrete-time models cf. [20, Chap. 1]. The other direction, from continuous-time dynamical systems to discrete-time dynamical systems finds applications in the study of periodic orbits, e.g., via *Poincaré return maps* [40]. Although we covered submanifold stabilization in Sect. 6.2, one can say more about the special case of periodic orbits, i.e., closed one-dimensional submanifolds of M. Here, we briefly highlight work due to Fuller. By exploiting the relation between the Euler characteristic, the Lefschetz number and homology, Fuller proved in 1952 the following.[1]

---

[1] Here, a *combinatorial manifold* is a topological manifold, with the atlas consisting out of homeomorphims that are piecewise linear (PL), also called *PL manifolds* [44, p. 4].

© The Author(s) 2023
W. Jongeneel and E. Moulay, *Topological Obstructions to Stability and Stabilization*,
SpringerBriefs in Control, Automation and Robotics,
https://doi.org/10.1007/978-3-031-30133-9_8

Theorem 8.1 (Fuller's condition for periodic orbits [19, Theorem 2]) *Let T be a homeomorphism of a combinatorial, compact, orientable manifold* M. *If* $\chi(M) \neq 0$ *then, T has a periodic point.*

Subsequently, Fuller provides a bound on the period [19, Theorem 3]. Theorem 8.1 has direct applications with respect to Poincaré return maps. If the domain of such a map has non-trivial Euler characteristic, a periodic orbit exists. Recall that for manifolds with zero Euler characteristic, the situation is less transparent, e.g., consider an irrational map on $\mathbb{S}^1$. See [7, 8] for more details and generalizations due to Byrnes.

8.2   Comments on Generalized Poincaré–Hopf Theory

The classical Poincaré–Hopf theorem is usually presented in its smooth form (Theorem 3.6), but as we already remarked, the same is true in the continuous (topological) setting (Corollary 3.1). Note that these results assumed M is boundary*less*, this can be relaxed. Assume $\partial M^m \neq \emptyset$, when $X \in \mathfrak{X}^r(M^m)$ points inward along ∂M, then the vector field indices sum up to $(-1)^m \chi(M^m)$. When the vector field points *outward* along the boundary, the Poincaré–Hopf theorem is unchanged [34, p. 35]. This has been recently exploited to assess uniqueness of equilibria [56, Theorem 1]. Generalizations to general vector fields being merely nonsingular on ∂M have been carried out by Morse [37] and Pugh [43]. Relaxing the compactness assumption on M has been studied in [11]. For a generalization applicable to hybrid systems, see [30].

8.3   A Decomposition Through Morse Theory

A smooth function $g : M^m \to \mathbb{R}$ is called a ***Morse function*** when all its critical points are nondegenerate. The ***Morse index*** of such a critical point $p \in M^m$ is the dimension of the subspace on which the Hessian of p is negative definite.[2] Then, the Morse Lemma [6, Lemma 8.2.4] states that around a critical point with Morse index k, there is a coordinate chart (U, φ) with $\varphi(p) = 0$ such that in coordinates $g(\varphi^{-1}(x)) = g(p) - x_1^2 - \cdots - x_k^2 + x_{k+1}^2 + \cdots + x_m^2$. Now, looking at a gradient vector field grad g, an equilibrium point with Morse index k has clearly vector field index $(-1)^k$. A powerful application is that one can extract a cell decomposition.

Theorem 8.2 (Morse indices and CW complex structures [6, Theorem 8.5.3]) *Let* $g : M^m \to \mathbb{R}$ *be a Morse function with* m_k *critical points of index k. Then,* M^m *is homotopy equivalent to a CW complex with* m_k *cells of dimension k, for* $k = 1, \ldots, m$.

As exploited in Theorem 4.2, this cell decomposition is a link to homology and $\chi(M^m)$. However, Theorem 8.2 is a manifestation of more general results, leading

---

[2] As discussed in Sect. 5.1, one does not need a connection to be able to perform the computation.

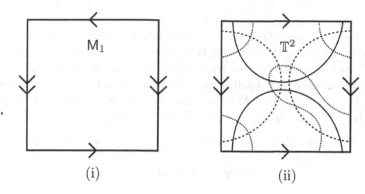

Fig. 8.1 Flat representations: (i) the Klein bottle (M_1 in Sect. 8.1); (ii) For $\mathbb{T}^2$ one sees via flat quotient representation $\mathbb{T}^2 = \mathbb{R}^2/\mathbb{Z}^2$ that $\mathsf{cat}(\mathbb{T}^2) \leq 2$

to the so-called *Morse inequalities* [28, Corollary 8.10.1]. In Sect. 8.5 we discuss a generalization of the work by Morse, largely due to Conley and Zehnder. Besides the original work by Morse [36] and generalizations due to Bott [2] and Smale [47], see also [33] and in particular [3] for more on Morse and his work

8.4   An Application of Lusternik–Schnirelmann Theory

Lyapunov functions $V : M \to \mathbb{R}_{\geq 0}$ can be used to capture qualitative behaviour of a vector field X on M. For example, the critical points of V can be related to equilibrium points of X. Based on the topology of M, the *Lusternik–Schnirelmann category* can be used to bound the minimal number of critical points any V can possibly have from below, thereby, bounding the number of equilibrium points of X.

Definition 8.1 (*Lusternik–Schnirelmann category* [14, Definition 1.1]) The category of a topological space X is the smallest $n \in \mathbb{N}_{\geq 0}$ such that there is an open covering $U_1, \ldots, U_{n+1}$ of X with each set U_i being contractible to a point in X. This number is denoted by $\mathsf{cat}(X) = n$ with the cover $\{U_i\}_{i \in [n+1]}$ being called categorical. If such a cover does not exist, $\mathsf{cat}(X) = +\infty$.

For the setting we consider, that of smooth or topological manifolds, one can consider a cover by open sets without loss of generality [14, Proposition 1.10].

To illustrate Definition 8.1, one can show that $\mathsf{cat}(\mathbb{R}^n) = 0$, $\mathsf{cat}(\mathbb{S}^n) = 1$ and $\mathsf{cat}(\mathbb{T}^n) = n$, see also Fig. 8.1 (ii). In contrast to $\chi(\cdot)$, $\mathsf{cat}(\cdot)$ is not trivial for compact Lie groups, e.g., $\mathsf{cat}(SO(3, \mathbb{R})) = 3$ and $\mathsf{cat}(SO(5, \mathbb{R})) = 8$.

For a more control-theoretic example, we return to the *switching controllers* from Sect. 7.3. If M is "*covered*" by local controllers all intended to asymptotically stabilize some point, their respective domains of attraction will be contractible by Theorem 6.4, so that one needs at least $\mathsf{cat}(M) + 1$ controllers.

Next we provide a corollary to the *Lusternik–Schnirelmann critical point theorem* [14] (for any smooth function over the compact smooth manifold M^n it holds that $cat(M^n) + 1 \leq crit(M^n)$ [14, 48]). For the precise definition of the *chain recurrent set* $\mathcal{R}(\varphi_X)$ of the flow φ_X we refer to [13, 26], but simply put, $\mathcal{R}(\varphi_X)$ is the intersection of attractor and repeller pairs under φ_X. Let A be an attractor, then, the corresponding repeller is $A^\circ = M \setminus \mathcal{D}(\varphi_X, A)$. Now the chain recurrent set is defined by $M \setminus \mathcal{R}(\varphi_X) = \cup_\alpha \mathcal{D}(\varphi_X, A_\alpha) \setminus A_\alpha$, or $\mathcal{R}(\varphi_X) = \cap_\alpha (A_\alpha \cup A_\alpha^\circ)$.

Corollary 8.1 (Critical points and limits sets) *Let M^n be a smooth, compact, boundaryless manifold. Then, if*

$$cat(M^n) + 1 > \chi(M^n), \tag{8.1}$$

there is no continuous vector field $X \in \mathfrak{X}^{r \geq 0}(M^n)$ with all of its equilibrium points $\{p_i^\star\}_{i \in \mathcal{I}}$ isolated and with $ind_{p_i^\star}(X) = (-1)^n$ such that $\mathcal{R}(\varphi_X) = \{p_i^\star\}_{i \in \mathcal{I}}$.

Proof For the sake of contradiction, there must be a smooth Lyapunov function $V : M^n \to \mathbb{R}$ that vanishes on $\mathcal{R}(\varphi_X) = \{p_i^\star\}_{i \in \mathcal{I}}$. Then, impose some metric $\langle \cdot, \cdot \rangle_p$ on $T_p M^n$, by assumption M^n admits at least one Riemannian metric. By bilinearity of the inner-product $\langle \cdot, \cdot \rangle_p$ on $T_p M^n$ the set $\{p \in M^n : grad\, V(p) = 0\}$ is invariant under the choice of metric (in contrast to neighbouring points, recall the example due to Takens [49, p. 231]). Then the claim follows directly from the Lusternik–Schnirelmann critical point theorem, Theorem 6.6 and [17, Corollary 2.3].

Although Corollary 8.1 is not surprising, one can study the gap $cat(M^n) - \chi(M^n)$ in order to get a better understanding of admissible multistable behaviour on M^n.

Example 8.1 (*Example* 6.5 *continued*) As $\chi(Gr(2, 3)) = 1$ and $cat(Gr(2, 3)) = 2$ [1, Theorem 1.2], Corollary 8.1 applies. Hence as in Example 6.5, we recover that although Theorem 6.6 applied, stability cannot be global indeed. Yet, as also mentioned in Example 6.5, if one assumes to have a asymptotically stable equilibrium point, then, there must be a non-trivial limit set as well.

8.5 Introduction to Conley Index Theory

The index theory as developed by Conley and coworkers enlarges the scope of Morse theory beyond non-degenerate critical points, in particular, beyond points.

Let us be given a dynamical system $(M, \mathbb{R}, \varphi)$. A **Morse decomposition** of a compact invariant set $A \subseteq M$ is a finite collection of compact disjoint invariant subsets $N_1, \ldots, N_n$ of A such that when $p \in A \setminus \sqcup_{j=1}^n N_j$ there are indices $i < j$ such that $\alpha(\varphi, p) \subset N_i$ and $\omega(\varphi, p) \subset N_j$. Such an order on $N_1, \ldots, N_n$ will be called *admissible*. Then, given a set $S \subseteq M$, let $I(S) = \{p \in S : \varphi(\mathbb{R}, p) \subseteq S\}$ denote the subset of S that is invariant under the flow φ. A compact set $N \subseteq M$ is called an **isolating neighbourhood** when $I(N) \subseteq int\, N$. Then, a set $S \subseteq M$ is an **isolated invariant set** if $S = I(N')$ for some isolated neighbourhood N' containing S.

Definition 8.2 (*Index pair*) Let S be an isolated invariant set. Then, the pair of compact sets (N, L) with $L \subseteq N$ is an *index pair* for S when

(i) $\mathrm{cl}(N \setminus L)$ is an isolating neighbourhood of S with $L \cap S = \emptyset$;
(ii) L is positively invariant in N, that is, if $p \in L$, then $\varphi([t_0, t_1], p) \subseteq N$ implies $\varphi([t_0, t_1], p) \subseteq L$;
(iii) L is an exit set on N for φ, that is, if $p \in N$ and there is a $t' > t_0$ such that $\varphi(t', p) \notin N$, then there is a $s \in [t_0, t']$ such that $\varphi(s, p) \in L$.

Now, let S be an isolated invariant set, with $N_1, \ldots, N_n$ an admissible Morse decomposition, then, there is a collection of subsets $M_0, M_1, \ldots, M_n$, called a *Morse filtration* such that $M_n \subseteq \cdots M_1 \subseteq M_0$ and whenever $i \leq j$, then (M_{i-1}, M_j) is an index pair for $N_{ij} = \{p \in S : \alpha(\varphi, p), \omega(\varphi, p) \subseteq N_i \cup N_{i+1} \cup \cdots \cup N_j\}$ [27, pp. 76–77]. Evidently, (M_0, M_n) is an index pair for S.

Then, the *cohomological Conley index* for an isolated invariant set S with index pair (N, L) is defined as $CH^{(\cdot)}(S) = H^{(\cdot)}(N, L)$. Similar to what was discussed in Chap. 6, the (co)homological Conley index can be used to provide for example necessary conditions for an equilibrium point to be asymptotically stable [13, Sect. I.4.3], [25, 38]. In this sense, $CH^{(\cdot)}(S)$ is analogous to the vector field index.

More generally, the **Conley index** can be defined as $h(S) = [(N/L, [L])]$, that is, as the homotopy type of the pointed space $(N/L, [L])$ [27, 46]. Although effectively intractable, in contrast to $CH^{(\cdot)}(S)$, the Conley index $h(S)$ allows for a result analogous to the Poincaré–Hopf theorem. To that end, define the *relative* Betti numbers as $b_k(B, C) = \mathrm{rank}\, H^k(B, C; \mathbb{Z}) = \mathrm{rank}\, H_k(B, C; \mathbb{Z})$. Additionally, set $p(t, B, C) = \sum_{k \geq 0} b_k(B, C) t^k$. If (B, C) is an index pair for some isolated invariant set A, with some abuse of notation put $p(t, h(A)) = p(t, B, C)$. Under this notation, the following is a generalization of earlier work due to Morse.

Theorem 8.3 (Conley index theorem [27, Corollary 2]) *Let $A \subseteq M$ be a compact, isolated, invariant set and $(N_1, \ldots, N_n)$ an admissible ordering of a Morse decomposition of A. Then,*

$$\sum_{j=1}^{n} p(t, h(N_j)) = p(t, h(A)) + (1+t)Q(t), \qquad (8.2)$$

noindent for $Q(t)$ a polynomial with non-negative integer coefficients.

We remark that $Q(t)$ depends on the chosen decomposition. Theorem 8.3 implies in particular that for a dynamical system $(M, \mathbb{R}, \varphi)$ over a compact manifold M one has

$$\chi(M) = \sum_{j=1}^{n} \sum_{k \geq 0} (-1)^k b_k(M_{j-1}, M_j) \qquad (8.3)$$

noindent for some Morse filtration $M_0, M_1, \ldots, M_n$. See that (8.3) is inherently *coarser* than (8.2), i.e., the excess term $Q(t)$ cancelled out. Also, when M is compact, the trivial decomposition $N_1 = M$ and the trivial filtration $(M, \emptyset)$ are always admissible.

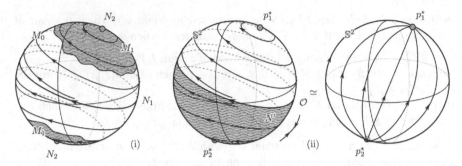

Fig. 8.2 (i) Example 8.2 with Morse decomposition (N_1, N_2) and Morse filtration (M_0, M_1, M_2); (ii) Example 8.2 with a coarse decomposition, see the equivalence through the lens of this decomposition. Also see that the rightmost figure is a typical instance where (8.1) holds with *equality*, that is, the vector field under consideration *does* exist here

Example 8.2 (*Admissible flows on* $\mathbb{S}^2$) Recall Fig. 6.4, we will apply Theorem 8.3. First, a Morse decomposition is given by (N_1, N_2) with orbit $N_1 = \mathcal{O}$ and equilibrium points $N_2 = \{p_1^\star\} \cup \{p_2^\star\}$. Then a Morse filtration $M_2 \subseteq M_1 \subseteq M_0$ is given by $M_0 = \mathbb{S}^2$, M_1 the disjoint union of sufficiently small compact spherical caps around $p_1^\star$ and $p_2^\star$ and $M_2 = \emptyset$. see Fig. 8.2i.

Only indicating non-trivial homology groups, as $H_k(M_0, \emptyset; \mathbb{Z}) \simeq \mathbb{Z}$ for $k \in \{0, 2\}$, $H_k(M_0, M_1; \mathbb{Z}) \simeq \mathbb{Z}$ for $k \in \{1, 2\}$ and $H_0(M_1, M_2; \mathbb{Z}) \simeq \mathbb{Z}^2$ we find that $Q(t) = 0$. Note, here we used the *wedge sum*, i.e., $M_0/M_1 \simeq \mathbb{S}^2 \vee \mathbb{S}^1$ [23, p. 10]. One could consider a coarser decomposition and ignore N_1, i.e., group $p_1^\star$, $p_2^\star$ and N' as Fig. 8.2ii. In that case, $Q(t) = 0$ and one effectively employs the Poincaré–Hopf theorem. Consult also the discussion in [27, Sect. 3.5].

Besides the original work by Conley, Zehnder and others [12, 13, 46], see [27, 35] for introductory works and for example [45] for a discrete-time analogue. Also, the work by Conley et al. builds upon that of Ważewski [54], [22, p. 280], which has been recently exploited in [41, 42] to provide further local obstruction theory.

8.6   Conclusion and Open Problems

This work aimed at providing an overview of how topology can provide for unique insights in control theoretic problems. This approach has a long and rich history and we believe this work shows it has more to offer for the future. In particular, Borsuk's retraction theory and the application of homotopy- and index theory akin to the framework by Krasnosel'skiĭ and Zabreĭko provide for a fruitful and unifying approach towards a mathematical control theory. We end with potential future work.

(i) **Sufficient conditions**: Controllability assumptions, the existence of a control-Lyapunov functions or structural properties like homogeneity can be exploited to provide *sufficient* conditions for certain feedback laws to exist. It would be interesting to see to what extent these results can be generalized to work for general control systems, e.g., $\Sigma = (M, \mathcal{U}, F)$, aiming at locally asymptotically stabilizing some closed set $A \subseteq M$. To that end, one needs to assert that A can be rendered invariant, there exists a dynamical system such that A is an attractor, locally, and so forth. Hence, sufficient conditions, especially conditions that lead to the construction of controllers are of great importance, yet, with the remarks of Casti [9] in mind, universal conditions might be too optimistic. Hence, a sub-question entails finding (more) appropriate classes of canonical control systems. An operator-theoretic start of this program is presented in [10].

(ii) **Numerical tools**: This line of work aims to provide necessary and sufficient conditions so that one can assess—*a priori*—if a stabilization problem has a solution. Clearly, these conditions must be easier to check than a brute force numerical experiment for this to be of practical value. In this regard, applying tools from computational homology [29] seems interesting. However, one might obtain corrupted information regarding the control system, e.g., F is based on experimental data. Recalling Remarks 6.1 and 6.5, computational tools must be able to provide certificates of accuracy, e.g., see [53] and references therein. Here, the field of *topological data analysis* (TDA) is also ought to play an increasingly important role. For instance, if one can sample points from Ω_t as in Theorem 6.3, TDA tools can be used to estimate if (6.2) is true.

(iii) **Other invariants**: Although finding the *"perfect"* invariant that allows for classifying dynamical systems up to conjugacy (Smale program) is too ambitious,[3] one might look for new invariants that capture (some) qualitative behaviour of interest. Related is the question, largely due to Conley, as posed in [31], can the homotopy from Theorem 6.13 be chosen to be *asymptotically* stabilizing throughout? Here, relaxing the notion of (uniform) asymptotic stabilization provides for many counterexamples cf. Fig. 3.4. We also believe that imposing restrictions on the class of Lyapunov functions to assert stability might be insightful. Moreover, it is not clear how to meaningfully extend this type of theory to generic input-output systems. In other words, can a similar program be outlined for *general* systems, going beyond dynamical systems cf. [21]. In particular, a program that allows for *compositional* thinking. Here we note that especially homology appears suitable, i.e., $(G_2 \circ G_1)_* = G_{2*} \circ G_{1*}$.

(iv) **Nonlinear system identification**: Our understanding of statistical finite-trajectory nonlinear system identification is improving, e.g., see [57], yet it is unclear how to go about identifying a nonlinear system (or equivalence class) globally. In particular, unstable equilibria are hard to identify. From Theorem 6.6 it follows that if one assumes that all unknown equilibria are asymptotically stable and the system is noiseless, then, one needs at least $\chi(M)$ trajectories to *"identify"*[4] these equilibria and their respective dynamics, locally.

---

[3] See for example [18] for related studies of the conjugacy problem in ergodic theory.

[4] Defining precisely what this means without resorting to questionable parametrizations is also largely open.

Combining the topological viewpoint with TDA and modern high-dimensional probability theory, e.g., [52], could be a fruitful combination. Other popular methods like neural networks, Gaussian processes, and so forth, all require further development of the theory. A promising direction exploits Koopman operator theory [5].

(v) *Feedback invariance*: As set forth in the monograph by Lewis [32], the lack of feedback-invariance present in most of the literature obstructs a principled control theory. In particular, the key necessary condition for control, that of controllability, is not invariant under feedback reparametrizations, see also early comments by Willems [55]. We also put emphasis on results that do not rely on the control system at hand, but merely on the underlying manifold M and the set A that is deemed to be stabilized, cf. Theorem 6.12. However, a variety of results, mostly contained in Sect. 6.1.1 *do* rely on the control system parametrization.

(vi) *Stochastic systems*: One can generalize Definition 5.5 to include disturbances [39, Definition 13.26]. Obstruction theoretic results that do include some form of noise often focus on moments [15, 16], however, how to extend the theory in the most meaningful way to stochastic systems is not clear.

(vii) *Zero Euler characteristic*: Set with zero Euler characteristic are inherently difficult to handle, cf. Theorems 6.5, 6.6, 6.9 and 6.11. As these sets are, however, abundant (odd-dimensional manifolds, Lie groups, periodic orbits), more theory is needed. Recalling Remark 6.4, retraction theory is of use, at the cost of being largely independent of the control system at hand. One could also look for other topological invariants that allow for a better classification of these sets.

More questions remain, conceptually, *why* are odd-dimensions seemingly inherently obstructive? What can be said about *finite*-time stabilization? How can we further integrate the structure of $\mathcal{U}$ and F in the analysis (beyond Theorem 6.14)? For instance, to get a better grip on Question iii. Most importantly, more topological obstructions are likely to be found. For example, one might use that for closed topological spaces X and Y it holds that $\chi(X \sqcup Y) = \chi(X) + \chi(Y)$, or by using that an orientable manifold M^n has even Euler characteristic if n is not a multiple of 4 [24, Theorem 1.2]. Or less recent, a closed connected smooth manifold M has $\chi(M) = 0$ if and only if it admits a smooth codimension-1 foliation [51, Theorem 1]. However, more impactful ought to be obstructions derived from Conley– and Morse–Bott theory or obstructions applicable in the context of robust- and hybrid control theory. In particular, theory well equipped to handle input-output systems. Although progress has been made with respect to obtaining a principled approach to stabilization on Lie groups [50] cf. Sect. 1.3, the central problem remains that of finding tractable sufficient conditions—as stressed throughout the literature [10, 31]. In particular, sufficient conditions that lead to implementable controllers. *Ce n'est pas tout.*

References

1. Berstein I (1976) On the Lusternik-Schnirelmann category of Grassmannians. Math Proc Cambridge 79(1):129–134
2. Bott R (1954) Nondegenerate critical manifolds. Ann Math 60(2):248–261
3. Bott R (1980) Marston Morse and his mathematical works. B Am Math Soc 3(3):907–950
4. Brin M, Stuck G (2002) Introduction to dynamical systems. Cambridge University Press, New York
5. Brunton SL, Budišić M, Kaiser E, Kutz JN (2022) Modern Koopman theory for dynamical systems. SIAM Rev 64(2):229–340
6. Burns K, Gidea M (2005) Differential geometry and topology: with a view to dynamical systems. Chapman and Hall/CRC, Boca Raton
7. Byrnes CI (2009) On the topology of Liapunov functions for dissipative periodic processes. Springer, Berlin, pp 125–139
8. Byrnes CI (2010) Topological methods for nonlinear oscillations. Not Am Math Soc 57(9):1080–1091
9. Casti JL (1982) Recent developments and future perspectives in nonlinear system theory. SIAM Rev 24(3):301–331
10. Christopherson BA, Mordukhovich BS, Jafari F (2022) Continuous feedback stabilization of nonlinear control systems by composition operators. ESAIM: Contr Optim Ca 28:1–22
11. Cima A, Manosas F, Villadelprat J (1998) A Poincaré-Hopf theorem for noncompact manifolds. Topology 37(2):261–277
12. Conley C, Zehnder E (1984) Morse-type index theory for flows and periodic solutions for Hamiltonian equations. Commun Pur Appl Math 37(2):207–253
13. Conley CC (1978) Isolated invariant sets and the Morse index. American Mathematical Society, Providence
14. Cornea O, Lupton G, Oprea J, Tanré D et al (2003) Lusternik-Schnirelmann category. American Mathematical Society, Providence
15. Elworthy K, Rosenberg S (1996) Homotopy and homology vanishing theorems and the stability of stochastic flows. Geom Funct Anal 6(1):51–78
16. Elworthy KD (1992) Stochastic flows on riemannian manifolds. Diffusion processes and related problems in analysis. volume II. Birkhäuser, Boston, pp 37–72
17. Fathi A, Pageault P (2019) Smoothing Lyapunov functions. T Am Math Soc 371(3):1677–1700
18. Foreman M, Rudolph DJ, Weiss B (2011) The conjugacy problem in ergodic theory. Ann Math 1529–1586
19. Fuller FB (1953) The existence of periodic points. Ann Math 229–230
20. Goebel R, Sanfelice RG, Teel AR (2012) Hybrid dynamical systems. In: Hybrid dynamical systems. Princeton University Press, Princeton
21. Golubitsky M, Guillemin V (1973) Stable mappings and their singularities. Springer, New York
22. Hartman P (2002) Ordinary differential equations. Society for Industrial and Applied Mathematics, Philadelphia
23. Hatcher A (2002) Algebraic topology. Cambridge University Press, Cambridge
24. Hoekzema RS (2020) Manifolds with odd euler characteristic and higher orientability. Int Math Res Notices 2020(14):4496–4511
25. Hui Q, Haddad WM (2006) Stability analysis of nonlinear dynamical systems using Conley index theory. In: Proceedings of the IEEE conference on decision control, pp 4241–4246
26. Hurley M (1995) Chain recurrence, semiflows, and gradients. J Differ Equ 7(3):437–456
27. Jost J (2005) Dynamical systems: examples of complex behaviour. Springer Science & Business Media, Berlin
28. Jost J (2017) Riemannian geometry and geometric analysis. Springer, Berlin
29. Kaczynski T, Mischaikow KM, Mrozek M (2004) Computational homology. Springer, New York

30. Kvalheim MD (2021) Poincaré-Hopf theorem for hybrid systems. arXiv e-prints. arXiv:2108.07434
31. Kvalheim MD (2023) Obstructions to asymptotic stabilization. SIAM J Control Optim 61(2):536–542
32. Lewis AD (2014) Tautological control systems. Springer, Cham
33. Milnor J (1963) Morse theory. Princeton University Press, Princeton
34. Milnor J (1965) Topology from the differentiable viewpoint. Princeton University Press, Princeton
35. Mischaikow K, Mrozek M (2002) Conley index. Handbook of dynamical systems, vol 2. Elsevier, Amsterdam, pp 393–460
36. Morse M (1925) Relations between the critical points of a real function of n independent variables. T Am Math Soc 27(3):345–396
37. Morse M (1929) Singular points of vector fields under general boundary conditions. Am J Math 51(2):165–178
38. Moulay E, Hui Q (2011) Conley index condition for asymptotic stability. Nonlinear Anal-Theor 74(13):4503–4510
39. Nijmeijer H, van der Schaft A (1990) Nonlinear dynamical control systems. Springer, New York
40. Parker TS, Chua LO (1989) Practical numerical algorithms for chaotic systems. Springer, New York
41. Polekhin I (2018) On topological obstructions to global stabilization of an inverted pendulum. Syst Control Lett 113:31–35
42. Polekhin I (2021) On the application of the Ważewski method to the problem of global stabilization. Syst Control Lett 153:104953
43. Pugh CC (1968) A generalized Poincaré index formula. Topology 7(3):217–226
44. Ranicki AA (ed) (1996) The Hauptvermutung book. Casson AJ, Sullivan DP, Armstrong MA, Rourke CP, Cooke GE. Springer, Dordrecht
45. Robbin JW, Salamon D (1988) Dynamical systems, shape theory and the Conley index. Ergod Theor Dyn Syst 8(Charles Conley Memorial Issue):375–393
46. Rybakowski KP, Zehnder E (1985) A Morse equation in Conley's index theory for semiflows on metric spaces. Ergod Theor Dyn Syst 5(1):123–143
47. Smale S (1967) Differentiable dynamical systems. B Am Math Soc 73(6):747–817
48. Takens F (1968) The minimal number of critical points of a function on a compact manifold and the Lusternik-Schnirelman category. Invent Math 6(3):197–244
49. Takens F (1971) A solution. In: Manifolds, Amsterdam. Springer, Berlin, pp 231–231
50. Teng S, Clark W, Bloch A, Vasudevan R, Ghaffari M (2022) Lie algebraic cost function design for control on Lie groups. In: Proceedings of the IEEE conference on decision control, pp 1867–1874
51. Thurston WP (1976) Existence of codimension-one foliations. Ann Math 104(2):249–268
52. Vershynin R (2018) High-dimensional probability: an introduction with applications in data science. Cambridge University Press, Cambridge
53. Vieira ER, Granados E, Sivaramakrishnan A, Gameiro M, Mischaikow K, Bekris KE (2023) Morse graphs: topological tools for analyzing the global dynamics of robot controllers. In: Algorithmic foundations of robotics XV. Springer, Cham, pp 436–453
54. Ważewski T (1948) Sur un principe topologique de l'examen de l'allure asymptotique des intégrales des équations différentielles ordinaires. In: Annales de la Société Polonaise de Mathématique
55. Willems JC (1998) Open dynamical systems and their control. Doc Math Extra ICM (III):697–706
56. Ye M, Liu J, Anderson BD, Cao M (2021) Applications of the Poincare-Hopf theorem: epidemic models and Lotka-Volterra systems. IEEE T Automat Contr
57. Ziemann IM, Sandberg H, Matni N (2022) Single trajectory nonparametric learning of nonlinear dynamics. In: PMLR conference on learning theory, pp 3333–3364

Series Editor Biographies

Tamer Başar is with the University of Illinois at Urbana-Champaign, where he holds the academic positions of Swanlund Endowed Chair, Center for Advanced Study (CAS) Professor of Electrical and Computer Engineering, Professor at the Coordinated Science Laboratory, Professor at the Information Trust Institute, and Affiliate Professor of Mechanical Science and Engineering. He is also the Director of the Center for Advanced Study-a position he has been holding since 2014 At Illinois, he has also served as Interim Dean of Engineering (2018) and Interim Director of the Beckman Institute for Advanced Science and Technology (2008–2010). He received the B.S.E.E. degree from Robert College, Istanbul, and the M.S., M.Phil., and Ph.D. degrees from Yale University. He has published extensively in systems, control, communications, networks, optimization, learning, and dynamic games, including books on non-cooperative dynamic game theory, robust control, network security, wireless and communication networks, and stochastic networks, and has current research interests that address fundamental issues in these areas along with applications in multi-agent systems, energy systems, social networks, cyber-physical systems, and pricing in networks.

In addition to his editorial involvement with these Briefs, Başar is also the Editor of two Birkhäuser series on Systems and Control: Foundations and Applications and Static and Dynamic Game Theory: Foundations and Applications, the Managing Editor of the Annals of the International Society of Dynamic Games (ISDG), and member of editorial and advisory boards of several international journals in control, wireless networks, and applied mathematics. Notably, he was also the Editor-in-Chief of Automatica between 2004 and 2014. He has received several awards and recognitions over the years, among which are the Medal of Science of Turkey (1993); Bode Lecture Prize (2004) of IEEE CSS; Quazza Medal (2005) of IFAC; Bellman Control Heritage Award (2006) of AACC; Isaacs Award (2010) of ISDG; Control Systems Technical Field Award of IEEE (2014); and a number of international honorary doctorates and professorships. He is a member of the US National Academy of Engineering, a Life Fellow of IEEE, Fellow of IFAC, and Fellow of SIAM. He has served as an IFAC Advisor (2017–), a Council Member of IFAC (2011–2014),

© The Author(s) 2023

W. Jongeneel and E. Moulay, *Topological Obstructions to Stability and Stabilization*,
SpringerBriefs in Control, Automation and Robotics,
https://doi.org/10.1007/978-3-031-30133-9

132   Series Editor Biographies

President of AACC (2010–2011), President of CSS (2000), and Founding President of ISDG (1990–1994).

Miroslav Krstic is Distinguished Professor of Mechanical and Aerospace Engineering, holds the Alspach Endowed Chair, and is the Founding Director of the Cymer Center for Control Systems and Dynamics at UC San Diego. He also serves as Senior Associate Vice Chancellor for Research at UCSD. As a graduate student, he won the UC Santa Barbara best dissertation award and student best paper awards at CDC and ACC. He has been elected as Fellow of IEEE, IFAC, ASME, SIAM, AAAS, IET (UK), AIAA (Assoc. Fellow), and as a foreign member of the Serbian Academy of Sciences and Arts and of the Academy of Engineering of Serbia. He has received the SIAM Reid Prize, ASME Oldenburger Medal, Nyquist Lecture Prize, Paynter Outstanding Investigator Award, Ragazzini Education Award, IFAC Nonlinear Control Systems Award, Chestnut textbook prize, Control Systems Society Distinguished Member Award, the PECASE, NSF Career, and ONR Young Investigator awards, the Schuck ('96 and '19) and Axelby paper prizes, and the first UCSD Research Award given to an engineer. He has also been awarded the Springer Visiting Professorship at UC Berkeley, the Distinguished Visiting Fellowship of the Royal Academy of Engineering, and the Invitation Fellowship of the Japan Society for the Promotion of Science. He serves as Editor-in-Chief of Systems and Control Letters and has been serving as Senior Editor for Automatica and IEEE Transactions on Automatic Control, as Editor of two Springer book series-Communications and Control Engineering and SpringerBriefs in Control, Automation and Robotics-and has served as Vice President for Technical Activities of the IEEE Control Systems Society and as Chair of the IEEE CSS Fellow Committee. He has coauthored 13 books on adaptive, nonlinear, and stochastic control, extremum seeking, control of PDE systems including turbulent flows, and control of delay systems.

Printed in the United States
by Baker & Taylor Publisher Services